Exploration and Mapping of the National Parks

Edited by
Jenny Marie Johnson

Occasional Paper No. 4

Map and Geography Round Table
of the
American Library Association

SPECULUM ORBIS PRESS
WINNETKA, ILLINOIS
1994

Library of Congress Catalog Card Number 93-80380
ISBN: 0-932757-04-9

Published by
Speculum Orbis Press
1050 Gage Street
Winnetka, Illinois
60093 USA
for the
Map and Geography Round Table
of the
American Library Association

CONTENTS

ILLUSTRATIONS

FOREWORD

Of a map's many uses for science, scholarship and travel, I have always valued the most romantic, its ability to induce vicarious travel. It is a dull heart, indeed, that is not whisked away to the place and time represented by a map. In an instant one can fly from home, office, or hospital room to share the Marquis de Rubi's view of Big Bend from Presidio del Norte, smell Yellowstone's sulfurous steam with John Colter, or hear Custer's last shots with Crazy Horse or Marcus Reno. Downtown Pittsburgh and Three Rivers Stadium can be gone and French and English colonials struggling again in the wilderness over Fort Duquesne.

The value of maps to the National Park Service is almost beyond estimation. The service holds in trust for present and future generations the greatest national treasures of history, nature, and scenic beauty; and through the National Register of Historic Places, Historic and National Landmarks, Rivers, Trails, and Recreation programs it helps and encourages others to follow its example. These are not abstractions that might be preserved in a book, a drawing, or a photograph. They are tangible matter: earth, water, stone, wood, and brick. Each is singularly and specifically located in its place on the face of the land and must be managed, preserved, protected, and interpreted in that place. The role of maps in these tasks is obvious. Virtually every visitors' center explains its resource through films, exhibits, and of course maps. When a visitor passes the entrance gate of a national park he or she receives information about how to understand and enjoy the resource without harming it and how to avoid being bitten by a snake, eaten by a bear, or otherwise injured. Included with this information, in every instance, is a map.

Modern maps can be useful research tools even for historians. Place names are durable, often outlasting the cultural context that gave them meaning. Cañon de las Linguas, the canyon of many tongues, reveals a place at the edge of the Staked Plains where Spanish-speaking Comancheros traded with tribes who spoke a variety of languages and, not infrequently, redeemed English-speaking captives in return for goods. But sometimes maps pose questions without answers. The Staked Plains itself, denominated Llano Estacado by Francisco Vasquez de Coronado, its first European explorer, leaves us wondering whether the Conquistador marked his trail by driving stakes into the ground or was reminded of a stockade as he approached the escarpment at its edge. Going-to-the-Sun Highway, as it scales the continental divide in Glacier National Park, might suggest a road into the sky but in fact it bears the name given by American Indians to a nearby mountain.

The greatest practical value of maps to the National Park Service people and their allies who struggle to preserve history and nature for the use and enjoyment of present and future generations is their utility in resource management. It is simply not possible to manage natural or cultural resources with any degree of confidence unless one first knows what the resources are and where they are located. For example, what are the number, the range, and the seasonal movement pattern of elk? What and where are the food plants on which they depend? What are the numbers and locations of other creatures that compete for that same food supply? Where are the historic structures and archaeological sites, and what is the nature and importance of each? Which topographic features are held sacred by neighboring American Indian tribes? A myriad of such questions must be answered before a spade of earth is turned to build a road or campground or other facility needed by visitors. If today's concept of a map could not extend to electronic patterns in microchips as well as to ink on paper, it would never be possible to manipulate the complex data needed by resource managers.

When the Secretary of the Interior launched a program in 1990 to preserve Civil War and other battlefields, and Congress created a special commission to study the problem, many of us dreamed of new statutory tools and economic incentives that would help protect the resources. However, the greatest single threat to battlefields comes from the lack of precise information about the location of important battle action and remaining battlefield features. Local officials who make the most important land use decisions usually know that a battle had occurred in the general vicinity of their community but almost never know the exact place where swords had crossed and blood had spilled. Using global positioning systems, which triangulate from satellites in space, the National Park Service staff have been able to reduce generalized information to highly specific data that can be understood on a human scale. Using geographic information systems, these precise data can be arrayed with other relevant data to plan where development can go without harm to resources and where it cannot. The resulting maps enable planners to know that moving a proposed road by a few feet may save a stone wall that had been assaulted by one set of heroes and defended by another or that a proposed warehouse would desecrate the spot where cavalry had charged cannons head-on. The maps even help to mobilize citizen interest in preservation and cause people to investigate the economic advantages of heritage tourism rather than intensive development.

It is with great pleasure that the American Library Association's Map and Geography Round Table presents this publication on the mapping of the national parks. Join me in reading of the first graphic recording of the Republic's crown jewels. Travel vicariously with the explorers and with the first readers of the maps to the wonders themselves. If you decide to plan a personal visit to a park, pull out a map and dream for a while about the places it shows. Look up unusual names in the encyclopedia. Go to the library and check out a book. Write for a park folder (and a map). A trip need not be confined to the actual travel.

JERRY L. ROGERS
Associate Director for Cultural
Resources and Keeper of the
National Register of Historic Places
United States National Park Service

INTRODUCTION

In 1987 the ALA Map and Geography Round Table (MAGERT) asked me to solicit papers focusing on the historical cartography of a representative group of the national parks with a geographical cross-section through the United States, although a western emphasis seemed inevitable. The overall intention of this proposal was creating a set of papers for presentation at our annual meetings in 1988 and 1989. All of the papers, including those not presented, were to be collected, edited, and submitted to the Round Table for publication in its Occasional Paper series.

Each author followed some general guidelines to achieve consistency. Maps of the parks, both early and contemporary, received major emphasis, and illustrations were encouraged for the presentations and published papers. Additionally, park legislative history, individual personalities, humorous anecdotes discovered, and general histories of geographical areas prior to park designations could be included.

The papers published here represent the culmination of the work of these authors, MAGERT's Publications Committee, and tireless editorial efforts of Jenny Marie Johnson. We certainly have a cross-section with the papers covering the first national park (Yellowstone) and a geographical expanse from Acadia to Hawaii. There are additional parks to be investigated, and it is hoped that this publication will rekindle interest in the mapping of our national parks, lead to other presentations, and maybe even another Occasional Paper.

DAVID A. COBB
Harvard University

"SUPREME AND DISTINCTIVE" ON THE EAST COAST: THE MAPPING OF ACADIA NATIONAL PARK

Nancy A. Kandoian

Map Division
The New York Public Library

When Acadia became a national park in February 1919, it was a historic occasion for the country and the Park Service as well as for the people who had worked to create the park. Lafayette National Park, as Acadia was then called, became the first national park east of the Mississippi River. It also became the United States' first coastal national park and the first that resulted from a gift of land to the government. What makes the park special even more than these firsts, what in fact made these premier positions possible, are its natural features.

The centerpiece of the park today, and the core of the park historically, is a range of mountains on Mount Desert Island, a unique formation east of Maine's mid-coastal region, where bald pink granite summits, uplifted by geologic forces, sculpted by glaciers and preserved by their resistant nature, look out over waters of the Atlantic Ocean, Blue Hill Bay, and Frenchman Bay. The highest of these summits, that of Cadillac Mountain at 1,530 feet above sea level, is the point of greatest elevation along the Atlantic coast of the United States. The park today extends beyond these mountains to the rocky eastern, western and southern shores of the island, eastward across Frenchman Bay to the mainland Schoodic Peninsula, and southwesterly to another island of prominent but lesser height, Isle au Haut.

The outstanding natural features of Mount Desert Island were so impressive that they were cited along with those of the Grand Canyon when Secretary of the Interior Franklin K. Lane made a "statement of national park policy" in a May 13, 1918, letter to Stephen T. Mather, Director of the National Park Service.[1] By this time, the area's historic significance and scientific interest already had been officially recognized with its designation as Sieur de Monts National Monument in 1916; its "scenery of supreme and distinctive quality" made it a natural for a national park according to the policies articulated in Lane's letter.[2]

EARLY MAPPING, TO 1800

Samuel de Champlain, sailing for the French in the early seventeenth century, explored, mapped, and described what is now the Maine coast. By this time, other explorers had already sighted the bare heights that

Champlain called "monts deserts." But the map makers who recorded these earlier explorers' discoveries had erroneously portrayed this island off the coast of Maine, separated only by a very narrow channel from the mainland, as part of the mainland.[3] Champlain determined the island nature of this mountainous coastal place.[4] After Champlain, French, Dutch, and eventually English maps of the northeast coast of North America routinely showed Mount Desert Island and displayed names derived from those that Champlain had aptly assigned to this and the other prominent island along this part of the coast. His carefully delineated 1607 map labels the islands "I. Monts desers" and "Isle haute."[5]

Though the scale of a map often did not seem to warrant the labelling of islands as small as these, their landmark heights made them worthy of mention on maps that would record discoveries and show information of interest to mariners. Thomas Jefferys made the point well, though somewhat later, on his *New Map of Nova Scotia, and Cape Breton Island with the adjacent parts of New England and Canada . . .*, 1775, scale ca. 1:1,900,000, in his *American Atlas . . .*, 1776, by noting next to Mount Desert Island, "seen 15 lgs. [leagues] off."[6] Essentially, the early maps of the Acadia National Park region were small scale maps of the coast of North America that, more or less, recorded discoveries and showed information of interest to mariners.

As history progressed, as needs and uses of the region developed, so did the scales of the maps increase and the types of maps specialize. Native Americans summered on Mount Desert Island, but it is not clear what type of mapping they may have done. The English and the French both, to varying degrees, were interested in settlement and trade along the coast of Maine. While French attempts at settlement were short-lived, permanent British colonial settlement began on the island in 1761 at the head of Somes Sound.[7] Not long after this, detailed mapping began in earnest. Protection of interests and resolution of conflicts meant military activity; that and trade translated into a need for nautical charts. Organized settlement created a need for maps showing property ownership. These are precisely the types of maps that resulted.

Massachusetts Governor Francis Bernard, recipient of the grant of Mount Desert Island by the colonial legislature, took surveyor John

Jones to survey the coast of the island.[8] Jones and Barnabas Mason completed a map of the whole island and its coastal surroundings, extending south and west to include Isle au Haut and east to include Schoodic Peninsula. As Morison notes, it provides one of the best early overall likenesses of the outline of Mount Desert.[9] Preserved in the Massachusetts Archives, the manuscript is inscribed *A Plan of the Islands Eastward Laying from Penobscot Bay & of the Granted Townships, with their Distances and Bearings from Each Other and from the Continent . . . Agreable to a Resolve of the Great and General Court for the Province of the Massachusetts Bay . . . February 13, 1765.* It shows not only names of islands and harbors but also what was probably Mount Desert Island's first road,[10] connecting the north coast of the island with the settlement at the north end of Somes Sound; the scale is one mile to the inch.

During the next decade, British military surveyors came to chart the area. Samuel Holland, as Surveyor General for the Northern District, surveyed the coast of New England; he "delineated all details and variations of the ground and shoreline."[11] Published products of Holland's efforts appear in Des Barres' *Atlantic Neptune.*[12] The section entitled "Charts of the coast and harbors of New England," published in 1781 "for use of the Royal Navy of Great Britain," includes four charts, dating from 1776 and 1781, ranging in scale from ca. 1:520,000 to ca. 1:49,000 showing the islands and peninsula that make up Acadia National Park today. As these were to serve as naval charts, the display of coastal features, structures along the coast, topographical features visible from the water, and obstructions to navigation was of primary importance. With hachures and, in some cases, a stippled kind of shading, these charts show a generalized hilly landscape on Mount Desert Island, Isle au Haut, and Schoodic Peninsula. Symbols for structures indicate that there were probably settlements by this time not only in the Somesville and Southwest Harbor areas, but also at Hulls Cove, Bar Harbor, Northeast Harbor and Bass Harbor, and on some of the surrounding islands as well. Rhumb lines connect shore points with the open sea for various compass headings, and soundings show depths in Somes Sound and in the thoroughfares between small islands in the vicinity; roads are not shown.

LANDOWNERSHIP MAPS

Settlement expanded on Mount Desert Island from the 1760s onward; early cadastral maps from the beginning of the nineteenth century show lots and property owners spreading out from the nuclei of first colonists on both the western and eastern halves of the island. Somes Sound physically divides the island, and the historical land grants had done the same thing.[13] Naturally enough, the earliest surviving property maps follow this division. Both are manuscript maps[14] at the scale of 80 rods to an inch (1:15,840) and are most readily accessible in the form of reproductions. The eastern half of the island is covered by *A Plan of the French Grant on Mount Desert . . . Copyed from a Plan Taken by John Peters, Esq., the Remainder of the Survey Taken by Me, James Peters, Bluehill, November 1807.* The western half of the island was mapped by a man named Salem Towne in 1808.[15]

When George B. Dorr, as Superintendent of Lafayette National Park, made his first annual report to the National Park Service, he described the significance of these two maps in showing the roots of the accepted property rights on the island.[16] In preparing the gift of land to be given to the federal government for the eventual formation of the park, Dorr's assistants searched titles to the land back to the original grants.[17] They found that these "two maps prepared by order of the Massachusetts court to show . . . accepted rights are still extant, though barely rescued from destruction." Referring to them as "the Peters plan" and "the Salem Town[e] plan" ("made by a bearer of that singular name"), he reported that they were "being photographed for the park to preserve their record and to show the island as it was at the beginning of its present occupation."[18]

From 1860 on, with the publication of Walling's map of Hancock County, Maine, the production of property maps of the Acadia National Park area paralleled similar commercial mapping throughout the eastern and midwestern United States. Like other landownership maps from the mid-nineteenth century, H. F. Walling's *Topographical Map of Hancock County, Maine, from Actual Surveys* is a large wall map with illustrations, insets, business directories, census figures, and a table of distances. At the scale of 1:70,000, the map shows topographical features, locations of houses, and names of property owners. By this time, not only was Mount Desert Island well-peopled (the census total for the three towns on the island, according to the map, was 3,930), but also settlers clustered along the northwest coast of Isle au Haut and at Mosquito Harbor on Schoodic Peninsula. Walling, working out of Boston and later New York, made and published maps of many towns and counties in New England and beyond.[19] It was not long before Maine mapmakers continued with work of similar genres, county and state atlases showing landowners and major topographical features, and listing local commercial establishments or including historical and descriptive text and statistical tables.

The *Atlas of Hancock County, Maine* was "compiled and published under the direction of Geo. N. Colby by S. F. Colby & Co." and "drawn . . . by H. E. Halfpenny & J. H. Stuart" in Ellsworth, Maine, in 1881. It includes landownership maps of the towns of Gouldsborough (containing Schoodic Peninsula) and Isle au Haut, and the four towns comprising Mount Desert and adjacent islands. Village plans of Bar Harbor, Mount Desert (Somesville), and Southwest Harbor show buildings and businesses, as well as property owners.

In 1887 George N. Colby of Houlton and Stuart of South Paris, Maine, published a large wall map of Mount Desert Island. Their *Map of Mount Desert Island, Sorrento, Portions of Lamoine, Hancock and Frenchmans Bay and Adjacent Islands, Maine*, scale ca. 1:17,500, locates landowners and topographical features as well as cultural features indicative of the activities of the time: lighthouses, a railway and United States Coast Survey station on Green Mountain (now known as Cadillac Mountain), and a ferry terminus.

By the turn of the century a wealthy summer community on Mount Desert Island was well established in and around Bar Harbor. A Summer Residents Association published an atlas that graphically displays the outlines of the so-called "cottagers" homes and the dimensions of their properties.[20] The atlas depicts, as well, other houses, buildings, and places of business in an area that extends north-south along the coast from Salsbury Cove to Anemone Cave, and east-west between one and two miles into the island from Frenchman Bay. This *Atlas of Bar Harbor and Vicinity in the Town of Eden, Hancock Co., Maine, 1904* is a record of the summer resort village and its outskirts on the eve of the movement to protect the natural glories of the island that resulted in the formation of the national park. Some of the cottagers, in fact, gave their land to the reservation that eventually became Acadia National Park. The association copyrighted the atlas in 1903 and published it at Bar Harbor; the Boston firm of Geo. H. Walker & Co. printed it.

NAUTICAL CHARTS

Many changes came to Mount Desert Island during the time real estate maps developed from manuscript sketches of settlers' lots to professionally printed atlases of a developed community with a service-based economy. Nautical charts had a heyday in that interval, when maritime activity was the vital part of the economy. When nautical charts came to be produced by the government, based on thorough and scientific surveys, accurately depicting the coastline and including topographical detail in the coastal region, they served an important role beyond their prime navigational function. They became the basis for future topographical and commercial mapping of the area. The official survey information was shared with the public, and private sector mapmakers and publishers used it regularly. They commonly cited the United States Coast Survey as the source for their maps; the citation lent authority to their products.

Guthorn[21] and McCorkle[22] discuss how privately made and printed charts filled a void until the official Survey of the Coast resulted in published charts in the mid-nineteenth century. An example of a private chart of the area of Acadia National Park (what Morison calls "the first general map of Maine following *The Atlantic Neptune* to depict the Island in any detail"[23]) was Captain Seward Porter's *Chart of the Coast of Maine*, 1837. Sheet 4 shows the southern part of Mount Desert Island and the western part of Schoodic Peninsula at the scale of ca. 1:63,360. Sheet 5, at ca. 1:180,000, shows "Isle Haut." Relief on Porter's charts is shown by hachures on land, and soundings show depths in the sound and in narrow thoroughfares between the smaller islands in the area. Some obstructions in the water are indicated, as are lighthouses. The only indications of settlement are at Southwest

Harbor at the southwest end of Somes Sound, in the locality of Somesville at the northwest end of Somes Sound, and on some of the smaller islands in the vicinity.

The painstaking activities of the United States Coast Survey reached northern New England by the 1850s. A triangulation station was set up on Mount Desert (i.e., on the top of Green Mountain) in 1856, though as early as 1849 a "primary unoccupied station" was designated on Mount Desert Island.[24] A. D. Bache, the Superintendent of the United States Coast Survey at that time, made a trip to Mount Desert Island to take readings at the station.[25]

The first large-scale chart of part of the park area was published in 1867; chart 317, *Winter Harbor, Maine*, depicts the harbor and the west coast of Schoodic Peninsula at the scale of 1:20,000. It shows harbor depths, nearby islands and lighthouses, and land features along the western side of the peninsula. Elevations are shown by contours; this seems to be the earliest example of such for Acadia National Park region maps.

In 1872 came the publication of chart 304, *South West Harbor and Somes Sound, Maine*, at the scale of 1:40,000. Its geographic coverage extends south and east of the harbor and sound to show much of the south coast of Mount Desert Island and the approaches to the island from the south through the Cranberry Isles. This was a high traffic area as Southwest Harbor was a busy fishing village, part of a shipbuilding town, and a port for steamboats that brought summer visitors to the island.[26] Though this chart cites "Topography by J. W. Donn . . . 1871," it shows no land detail, neither hachures nor contours for relief, and no interior place names or other culture features, excepting lighthouses.

The Coast Survey issued its topographic tour de force for the whole island of Mount Desert in October 1875 with the publication of its chart 103(2) at the scale of 1:80,000 (fig. 1). Listed in the chart catalog as a topographical map,[27] it is exactly that. No hydrographic information is given, so the island, with relief shown by hachures enhanced by a shading effect, sits in a sea of blank paper. It forms a striking image, alone, without mainland or soundings of surrounding waters, with only a few of the closest neighboring islands. The fine engraving of the hachures and other symbols for landscape and vegetation features, and

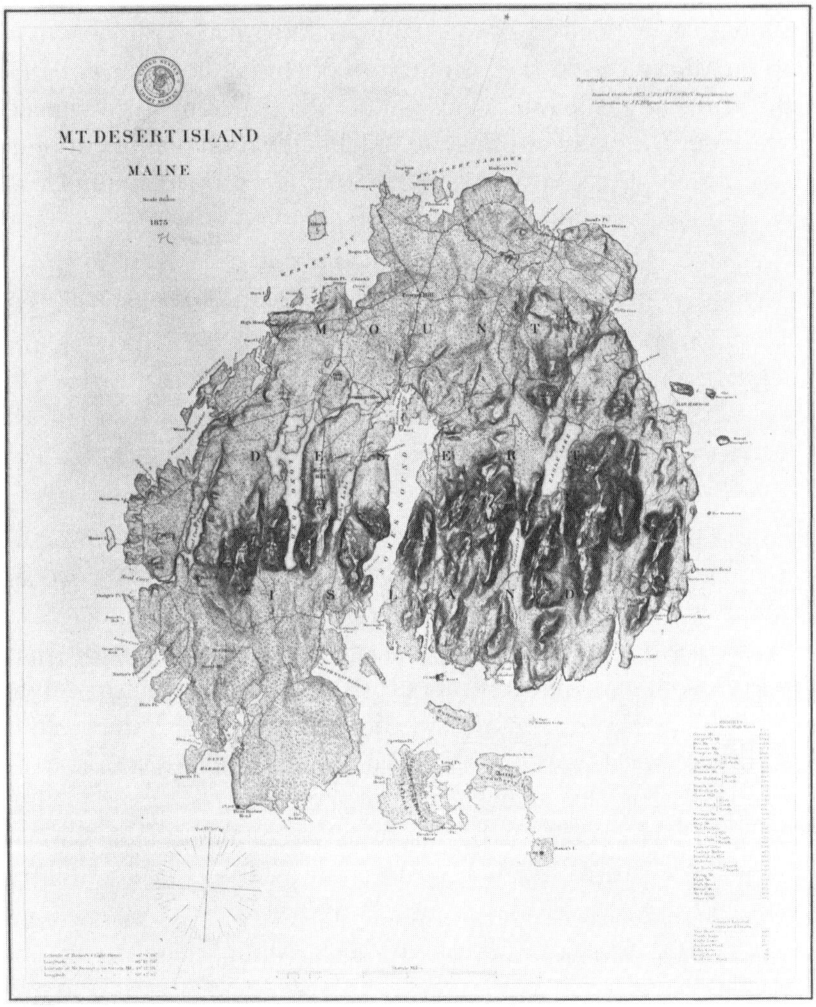

Fig. 1. Courtesy of Map Division, The New York Public Library, Astor, Lenox & Tilden Foundations, The Research Libraries.

the orientation of the map with north toward the upper right, makes a dramatic statement that a range of mountains, superior to other rises in the vicinity, crosses the island at a nearly perpendicular angle to Somes Sound. Placenames, roads, houses, surface water features, and lighthouses are shown, and figures for elevations are given in a list in the lower right corner of the map. There is a triangular symbol on the summit of Green Mountain, presumably showing the location of the triangulation station. The annual reports of the Coast Survey explain that, in the survey's well-ordered scheme of activities, topographic surveying preceded hydrographic sounding. The accuracy of the latter depended on the careful completion of the former.[28] The 1875 Coast Survey topographical map was, then, an intermediate dividend in the process that moved toward the goal of nautical chart production.

In 1882 a combination topographic map/hydrographic chart of Mount Desert Island was published by the recently renamed Coast and Geodetic Survey; the scale for chart 292 is 1:40,000. Relief is indicated by contours, a first for the island as a whole. Soundings and some offshore stippling show depths in the waters all around the island, and many surrounding islands and the mainland context are included.

During the 1880s, the Coast and Geodetic Survey produced a number of other charts, coast and harbor charts at scales of 1:80,000 and larger, for the mid- to eastern segment of the Maine coast. In 1883, a 1:80,000 scale chart of *Mount Desert Island, Frenchman's and Blue Hill Bays, and Approaches* (chart 103) was issued. Its coverage extends eastward to include Schoodic Peninsula. Separate charts of the eastern and western parts of Mount Desert Island (charts 306 and 307, respectively) at the scale of 1:40,000 were published in 1885. Chart 306 includes the part of Schoodic Peninsula that is now a component of Acadia National Park. A Bar Harbor chart (chart 318) at the scale of 1:10,000 was issued in 1885. Isle au Haut had been covered at 1:80,000 on chart 104 (Penobscot Bay), issued in 1880; and at 1:40,000 on chart no. 309 (East Penobscot Bay), issued in 1882.[29] Topographic features continued to be included along with hydrographic information on these charts.

Chart catalogs show that while some of the original charts, such as 103$^{(2)}$ and 292, were discontinued from publication during the next few decades, basic coverage for the area continued with new editions, published at varying intervals, of most of the 1:40,000 and 1:80,000 charts.[30] The larger scale harbor charts were reprinted with corrections from time to time.

PRE-PARK TOURIST MAPS

After American artists of the Hudson River School began to visit Mount Desert Island in the 1840s, the beauties of the island were introduced to eastern city-dwellers through works of art.[31] Steamboat service to the island began in the 1850s and expanded after the Civil War;[32] railroad lines made connections with the steamers, and one line eventually extended to a terminus within a ferry ride of Mount Desert Island. These conveyances brought ever larger numbers of urbanites who could afford the time and expense of travelling and spending summer vacations on the coast of Maine. In response to the needs of a new group of people relating to the land and sea of Mount Desert in a different way, another type of map of the area became prevalent. Tourist maps, small folding maps showing the general features of the island, places of interest and ways to get around, were included in guidebooks to Mount Desert published in the 1870s and 1880s and were distributed by the railroad and steamship companies that served the island.

Maps are included in some editions of Clara Barnes Martin's guidebook, *Mount Desert on the Coast of Maine.* Published in Portland by Loring, Short & Harmon, editions appeared from at least 1870 to 1885, building on a privately printed work from 1867.[33] A folded map, *Somes Sound, Mt. Desert Isd.: an Outline Sketch,* showing mountain names, spot heights, roads, buildings, and settlements is found inside the front cover of the 1874 edition, and *Eastern Part of Mt. Desert I.: an Outline Sketch Reduced from the Original Topographical Sheet, John W. Donn, Assistant, U.S. Coast Survey* is inside the back cover. On the list of illustrations within the guidebook, Martin credits the survey as the source of both maps:

> The first of these maps has been copied from the chart issued by the Coast Survey; the other has been prepared from unpublished

drawings by the courtesy of Mr. John W. Donn, of the Coast Survey, under permission from Prof. Peirce, the recent superintendent.[34]

The publication of Martin's 1874 guidebook predated the Coast Survey's publication of a chart detailing the eastern part of Mount Desert Island; but the credit statement suggests that the Coast Survey made its drawings available to those ready to publish maps.[35] Martin's second map shows spot heights, roads, trails, buildings, and placenames. The 1877, 1880, and 1885 editions each include a reduced facsimile of the 1875 Coast Survey topographical map of Mount Desert Island (chart $103^{(2)}$) folded and bound into the book.

The steamship and railroad lines that served Mount Desert promoted their destinations in the process of promoting their services. Separately issued maps and maps on the versos of timetables advertised the sinuous coast of Maine and the mountainous terrain of Mount Desert Island to which one could travel on a combination of train and steamer. Coast Survey charts probably provided a base for most of these maps, and in at least one case, where the map bears an obvious resemblance to the Coast Survey topographical map of 1875, the source is cited. *Mt. Desert Island, Maine, from U.S. Coast Survey,* "issued by Passenger Dep't Eastern R.R. Co.," undated but probably from the late 1870s, is uncolored, shows relief with hachures, includes only the closest neighboring islands in addition to Mount Desert, shows roads, houses, and placenames, and separately lists elevations of mountains, cliffs, lakes and ponds. A triangle without labelling marks the Coast Survey station on the summit of Green Mountain. The scale is reduced from that of the Coast Survey map, and the addition of a bold, dashed line shows the "route of steamers connecting with Eastern R.R. for Boston."

Maps distributed by the Eastern Railroad, the Boston & Maine Railroad, and the Boston & Bangor Steamship Company from the 1880s to the early 1900s show more of Frenchman, Blue Hill, and Penobscot bays, including some or all of Isle au Haut and Schoodic Peninsula; they show later developments on Mount Desert Island, such as the steamer on Eagle Lake and railway up Green Mountain; and they show more transportation routes and connections.

PATH MAPS

Mount Desert Island historians describe a succession in the type of summer visitors who went to the island in the mid- to late nineteenth century. After the artists came scientists, who found opportunities for the study of wildlife and geology on the island, and professional people (ministers, lawyers, college presidents) and others who could manage long summer vacations.[36]

Edward L. Rand fit both of these descriptions. He was a lawyer in Boston, and in addition, he had developed an interest in botany. He pursued this interest not only as a student camping on the island but also as an adult who undertook his avocation with the rigorousness of a professional.[37] His work observing and describing plants, eventually in collaboration with John H. Redfield, culminated with the publication in 1894 of *Flora of Mount Desert Island, Maine: A Preliminary Catalogue of the Plants Growing on Mount Desert and the Adjacent Islands.* Rand and Redfield's monograph is accompanied by "a geological introduction by William Morris Davis and a new map of Mount Desert Island" that Rand himself compiled.[38]

In their introduction, Rand and Redfield describe the rationale for the map; they wanted to indicate "stations for the various Island plants," and "it was necessary in all cases to give a name to the station that should be both accurate and well known as a matter of geographical nomenclature."[39] Along with this concern, a clear exasperation with existing maps gave Rand the impetus for making the map, an exasperation expressed about "the two maps most readily consulted, the Land Map of Colby and Stuart [1887] and the Coast Survey Map [1875 or 1882]," in the introduction:

> . . . the geographical nomenclature employed on the Coast Survey Map, and followed in some degree on the Land Map, is often, we regret to say, absolutely erroneous. In many a case, indeed, there is no explanation whatever to account for the blunders, except that the officers in charge of the work must have coined names for their own use, regardless or in ignorance of the fact that there might be names already attached to the places in question. In other cases, by some

curious mistake, names have been carelessly transposed and interchanged.[40]

They offer an example: "A very striking instance . . . is found in the unauthorized use of Turtle Lake for Bubble Pond, or for the oldest name of all, now obsolete, Southeast Pond."[41]

So with "much careful investigation of ancient maps, plans, and records," "the invaluable assistance [of] natives of the Island who knew and loved it well,"[42] and "topography adapted from the United States Coast Survey" (as stated on the map itself), Rand created his *Map of Mount Desert Island, Maine.* A contour map at the scale of 1:40,000, it follows conventions of United States Geological Survey maps in style and color, at a time when the USGS had not yet done detailed mapping of the area: sea waters are blue, contour lines are brown, and culture features are black. (One confusing characteristic, however, is that streams are also black.) Rand gives credit on the map itself to the Coast Survey for topography and to "trustworthy authorities" for nomenclature; he elaborates on these credits in the text of the book's introduction. It is fitting that he counters his criticism of the Coast Survey's haphazard attention to placenames with applause for their basic surveying and mapping of physical features:

> By kind permission of Prof. Thomas C. Mendenhall, Superintendent of the United States Coast Survey [by this time, in fact, Coast and Geodetic Survey], we have used the Coast Survey Map for the important physical features, making here and there a few corrections, and supplying a few omissions. We wish here to express our appreciation of his courtesy, which has enabled us to give a much better map to the public.[43]

Yes, "the public." Rand goes on to indicate that he has in mind not just the wildflower-collecting public, but a wider audience for his map:

> New roads, the town boundaries, and additional wood roads and paths have been added, the different post-offices indicated, and such points of interest named as it seemed would make a map not only suitable for our purpose, but of value to any one interested in the Island.[44]

With a view to publishing future editions, and revealing some of the pride they felt in Rand's accomplishment thus far, they solicit "other information that [would] serve to make the map more perfect and useful to those interested in the Island."[45]

Rand subsequently joined with two other summer residents of Mount Desert Island to correct and update the map, and to publish a derivative map at a larger scale. They did this for the general public and particularly for those who practiced what had become favorite recreational activities on the island in summer…walking, hiking, and driving (i.e., riding with horse and carriage). Rand's colleagues in this endeavor were Waldron Bates, like Rand a Boston lawyer, and Herbert Jaques, a Boston architect.

Beginning in 1896, Bates, Rand, and Jaques published the *Path Map of the Eastern Part of Mount Desert Island, Maine*, at scale 1:25,000, and a new edition of the *Map of Mount Desert Island, Maine*, at 1:40,000. These maps use contours to show relief, and their coloring follows that of the original Rand map, with the improvement that streams are shown in blue. On the path map, "paths and trails are shown by broken red lines. Red numbers indicate distances in miles." On the maps of the entire island, paths are shown but are not highlighted in red; they are black, like the roads.

Sources are cited on the 1896 maps in this way:

> Topography and hydrography adapted from the Map of Mount Desert Island issued by the United States Coast and Geodetic Survey, June 1882. Nomenclature…revised and corrected…adapted from the map dated June, 1893, compiled for the Flora of Mount Desert Island.

Revisions were made to the path map in 1900 by the addition of new paths in a darker red, by hand, and by the stamping of "with new paths 1900" onto the cartouche. New editions of the map, with scale reduced to 1:29,000, were published beginning in 1901, at varying intervals (fig. 2). New editions of the Bates, Rand, and Jaques map of the whole island, without the highlighting of the paths, were published beginning in 1906.

A noteworthy change to the Bates, Rand, and Jaques maps over time is the addition of green coloring, signaling the coming of the park. Beginning in 1909, green appears on the land mass of the island and is explained as "reservations." The green grows in area and designates

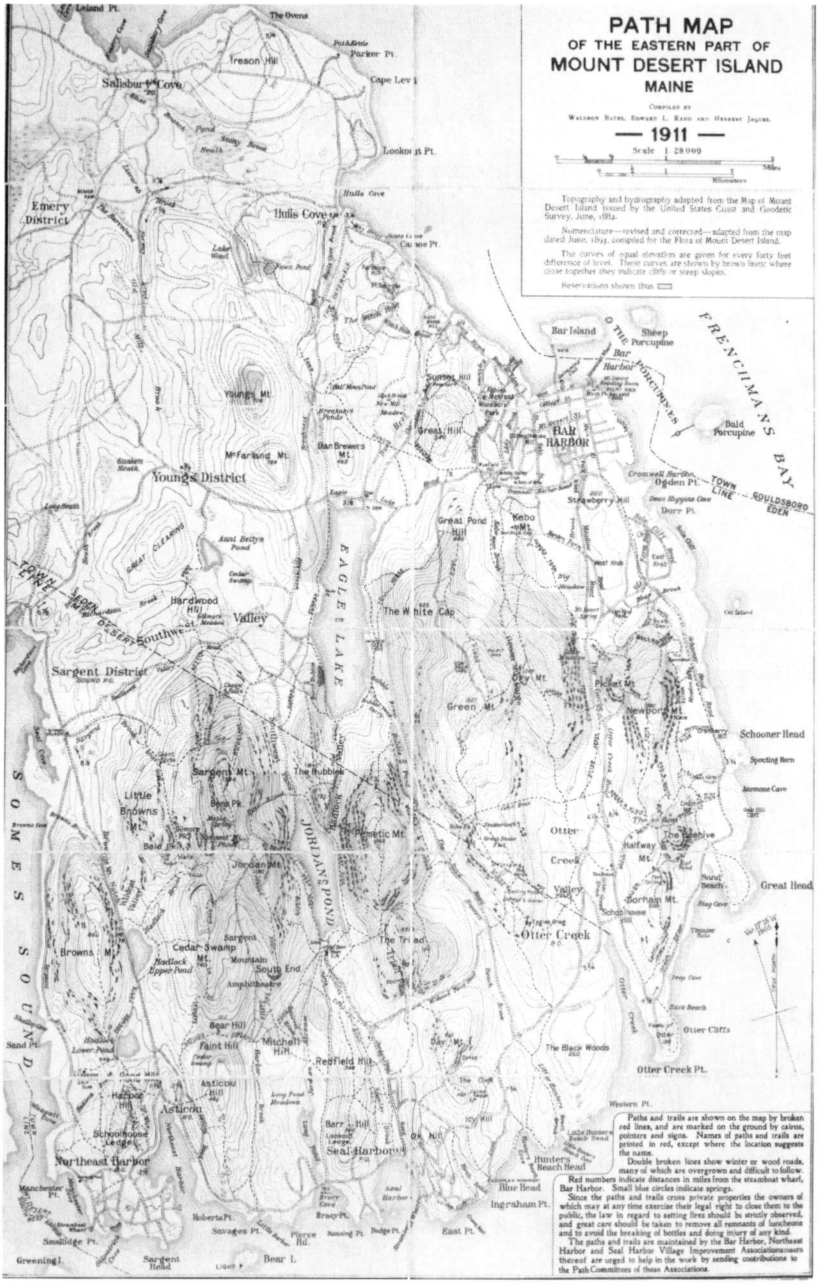

Fig. 2. Courtesy of Map Division, The New York Public Library, Astor, Lenox & Tilden Foundations, The Research Libraries.

"United States Government Monument," with green stripes showing "other public reservations," on the 1917 map. By this time, Sieur de Monts National Monument, the forerunner of Lafayette National Park, had been established. The green becomes "Lafayette National Park and other public reservations," "The area of Lafayette National Park and other reservations (so far as disclosed)," and finally "Acadia National Park and other reservations (so far as disclosed)" on later editions. The story behind these changes is the story of the growth of Acadia National Park.

UNITED STATES GEOLOGICAL SURVEY MAPS

After Bates, Rand, and Jaques had published their first map together, the United States Geological Survey (USGS) began to map the area that eventually became the park. From 1901 to 1903, surveys were carried out to supplement or update information from the Coast and Geodetic Survey. Fifteen-minute topographic quadrangles, part of the national series of maps at the scale of 1:62,500, were published in 1904 for sections called *Bar Harbor* (covering eastern Mount Desert Island and Schoodic Peninsula), *Mount Desert* (covering western Mount Desert Island), Swan Island (covering southern Mount Desert Island), and *Deer Isle* (covering Isle au Haut). Reprints with changes were issued in 1909 for *Bar Harbor*, 1910 for *Mount Desert*, and 1912 for *Swan Island* and *Deer Isle*.

While geological surveying had begun in Maine in the 1830s with the work of Charles T. Jackson for the state,[46] the first geological maps specifically of Mount Desert Island were published by the Geological Survey in 1889. Nathaniel Southgate Shaler carried out geological studies and produced two geological maps that were published as part of the *Eighth Annual Report of the United States Geological Survey . . . 1886-'87*. The maps are titled *Quaternary Deposits of Mount Desert Island, Maine* and *Geologic Map of Mount Desert Island, Maine*, and the scale of each is 1:100,000.[47]

Isle au Haut was mapped geologically (and topographically, as well) at the scale of 1:125,000 in the *Penobscot Bay Folio, Maine*, number 149, of the *Geologic Atlas of the United States*. The U.S. Geological Survey published this folio in 1907. Mount Desert Island and Schoodic Peninsula were not included in the *Geologic Atlas* folio series.

ROAD MAPS

While roads appeared on maps of Mount Desert Island from the 1760s onward, a genre of road maps that not only showed labeled roads but also distinguished roads of different qualities and different uses began to evolve with the approach of the twentieth century. The Bates, Rand, and Jaques maps titled *Map of Mount Desert Island, Maine* (1896+) are more than road maps because of their contours; yet they might be considered road maps nonetheless. Some roads are labelled, and more than one symbol represents roads (double solid lines and double dashed lines) although no legend explains the different symbols. The passing years that brought new editions of this map also brought the beginnings of private use of the automobile, disputes about whether to allow automobiles on the island, and controversies about the building of new roads.[48]

A 1911 *Map of Mount Desert Island, Maine*, drawn by W. G. Bunker with copyright by W. H. Sherman, is a road map of sorts.[49] Showing relief with hachures and spot heights, it distinguishes "roads and drives," "winter or wood roads," and "foot paths." It also labels many of the roads and presents the route of a "proposed auto road to Bar Harbor." This proposed auto road was a compromise offered in the struggle between factions in favor of and against the use of automobiles on the island; limited auto use along a road providing access between the mainland and the main resort village on the island was a gradual way to introduce cars.[50] The proposed road was not built as shown on the Bunker-Sherman map but the town did open to automobiles in 1913, as did the rest of the island in 1915.[51]

The Bates, Rand, and Jaques map of Mount Desert Island was first titled a "road map" with the 1934 edition. This map includes a legend and distinguishes "main motor roads," "other roads open to motors," "Rockefeller carriage roads," and "private and wood roads." John D. Rockefeller, Jr., who summered at Seal Harbor on Mount Desert Island,[52] put in order and expanded a network of roads for horse and carriage travel on his own and neighboring properties. The Rockefeller plan and the work on these roads by carefully selected engineers, landscape architects, architects, and construction people was carried on from 1913 to 1940.[53] During these decades, Sieur de Monts National Monument was established, the monument became a national park, and Rockefeller gave much of his land, including carriage roads, to the park. The carriage roads have become one of the special features of Acadia National Park and a special feature shown on maps of the park.

SIEUR DE MONTS NATIONAL MONUMENT

President Woodrow Wilson signed a proclamation declaring the creation of Sieur de Monts National Monument on Mount Desert Island, Hancock County, Maine, on July 8, 1916 (39 Stat. 1785-1791).[54] The monument was created from land given to the federal government by a group known as the Hancock County Trustees of Public Reservations. The gift and proclamation were the culmination of efforts by the trustees, chartered in 1903, to acquire, hold, and maintain lands for public use "by reason of scenic beauty, historical interest, sanitary advantage, or other like reasons."[55] George B. Dorr, a Bostonian and summer resident of Bar Harbor, was the prime mover among the Hancock County Trustees, and his memoirs tell the anecdotal and political history of the efforts.[56]

A detailed description of the boundaries of the approximately 5,000 acres that made up the gift of the Hancock County Trustees of Public Reservations, the land that became the national monument, is included in the proclamation along with a small (scale ca. 1:70,000), uncolored map of the area (39 Stat. 1785, facing page) (fig. 3). The map delineates the boundaries of the monument and shows some well known landmarks...Green Mountain, Eagle Lake, Jordan Pond and part of the Frenchman Bay coast...but without the frame of reference of the whole island of Mount Desert. The Green Mountain summit is within the monument boundaries, and the title of the map calls attention to this fact: *Sieur de Monts National Monument, Mount Desert Island, Maine, Embracing the Island Summit and about Five Thousand Acres of Adjacent Lands.*

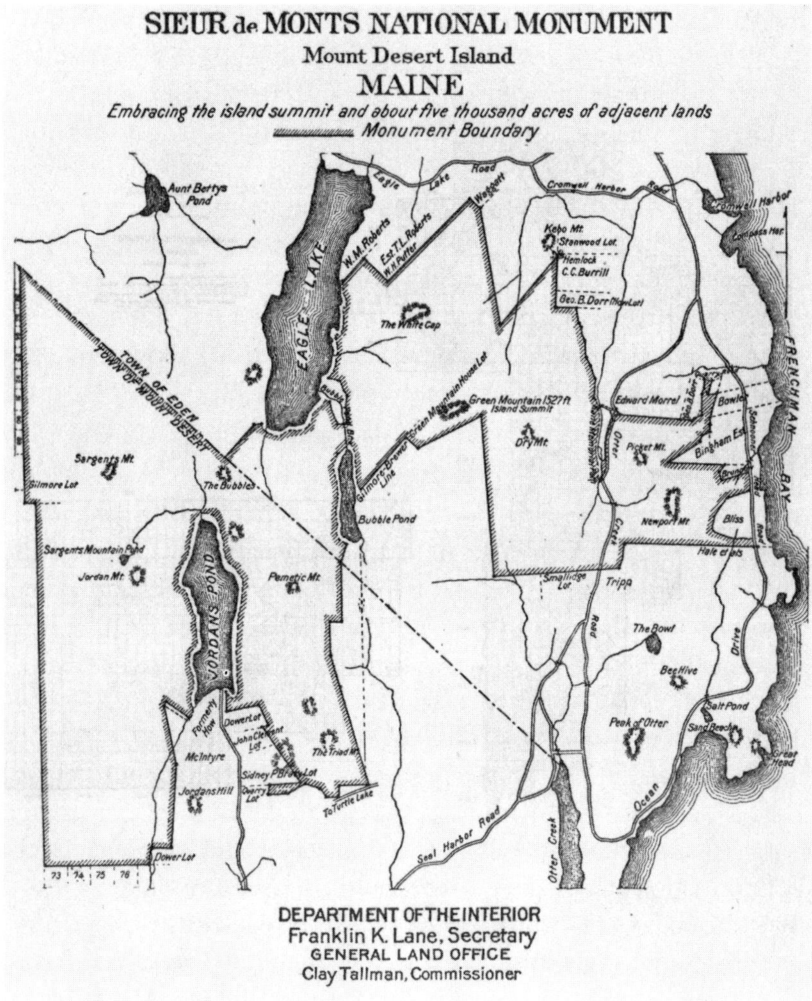

Fig. 3. Courtesy of Economic and Public Affairs Division, The New York Public Library, Astor, Lenox & Tilden Foundations, The Research Libraries.

Among the few maps published during the Sieur de Monts National Monument period an important discrepancy exists. The Geological Survey fifteen-minute topographic quadrangles available for the area at the time (*Bar Harbor* quadrangle, edition of April 1904, reprinted May 1909, and *Mount Desert* quadrangle, edition of 1904, reprinted May 1910 with corrections) were overprinted in red with the boundary and name of the national monument and the margin notation "National monument added Sept., 1916." The *Bar Harbor* quadrangle shows a national monument boundary that puts the summit of Green Mountain, the "island summit," **outside** of the new national monument. Editions of the Bates, Rand, and Jaques *Map of Mount Desert Island, Maine*, scale 1:40,000, and *Path Map of the Eastern Part of Mount Desert Island, Maine*, scale 1:29,000, published in 1917, show the Green Mountain summit **within** the monument.

It is not clear that the Green Mountain summit discrepancy has ever been publicly recognized or resolved. Nevertheless, further evidence supports the position that the summit was within the original national monument. There is specific mention of the Hancock County Trustees' acquisition of the "summit tract" of Green Mountain in 1908 in Dorr's book[57] and in *The Hancock County Trustees of Public Reservations: An Historical Sketch and a Record of the Holdings of the Trustees.*[58] *Hancock County Trustees of Public Reservations Gifts of Land for Acadia National Park, Mount Desert Island, Maine*, at the scale of ca. 1:60,000, accompanies the *Historical Sketch* and lays out the pattern of land parcels acquired by the trustees and donated for the monument and park. A key to the map tells that "Lot no. 11" is "the 'Mountain House Lot' consisting of 85.4 acres at the summit of Green Mountain." It adds that the deeds were recorded in 1908 and conveyed to the United States under the same registry numbers as other gifts that were clearly (i.e. on all maps) part of the original national monument.[59] The map was compiled by Charles W. Eliot, 2d, in 1935 and produced by The National Survey, Chester, Vermont.

Among the projects which Dorr pursued during the national monument years, while he served as custodian of the monument, was the publication of a series of illustrated papers on various topics related to the natural life, environment, and early historical background of the area; these were the "Sieur de Monts Publications."[60] There were no large scale maps among the Sieur de Monts Publications; yet, another kind of map of Mount Desert Island made an appearance twice: a photograph of a three-dimensional relief model appears in *The Coastal Setting, Rocks, and Woods of the Sieur de Monts National Monument* and in *The Sieur de Monts National Monument and the Wild Gardens of Acadia*, numbers 4 and 22 in the series.[61]

Dr. Robert Abbe, a New York surgeon who pioneered the use of radium to treat cancer in the United States,[62] was another summer resident of Bar Harbor who, through his serious pursuit of avocation, brought developments to the mapping of Mount Desert Island. He built relief models of the island which survive in the park and at libraries, historical societies, and museums on the island. The one photographed for the Sieur de Monts Publications is titled *The Beautiful Island of Mount Desert* and was apparently made for the Building of Arts, an arts center in Bar Harbor.[63] The relief models are a most appropriate medium for showing the island's glacial geomorphology. At least two of Abbe's models were copyrighted in 1916. Their horizontal scale is approximately 1:40,000. Mountains are named, roads and paths shown, and lighthouses located.

ESTABLISHMENT OF THE PARK AND USGS MAPS

Sieur de Monts National Monument became a national park, initially named Lafayette National Park, when President Wilson signed Congress' legislation creating the park on February 26, 1919. P.L. 65-278 was printed in the statutes (40 Stat. 1178) without an illustrative map, but the law referred to the same lands that had become Sieur de Monts National Monument by presidential proclamation on July 8, 1916.

In George B. Dorr's first annual report as superintendent of Lafayette National Park in 1919, he remarked that:

> One of the most pressing needs of the park is that of a good survey and map. The old Government survey of Mount Desert Island was made long since and hastily, with numerous inaccuracies. Every visitor to the park desires and needs a map for guidance . . .[64]

The first Geological Survey map of Lafayette National Park as a whole was published in 1922 but it was based on the earlier surveys. This map portrays the entire island of Mount Desert "from U.S. Geological Survey topographic maps of Mt. Desert, Bar Harbor, and Swan Island quadrangles, surveyed in 1901 and 1902," at the same scale as those fifteen-minute quadrangles, 1:62,500. The park at this time was still exclusively on the east side of the island, east of Somes Sound. The map shows the boundaries to be the same as the original national monument boundaries, but with the continued discrepancy in the area of the island summit, formerly Green and by this time named Cadillac Mountain. Like the USGS topographic quadrangles with 1916 red overprinting, the 1922 USGS map excludes the summit from the park. The USGS map of Lafayette National Park, edition of 1922 reprinted with changes in 1928, exhibits only a small park boundary change, plus road differences from the 1922 map. There is an addition to the park at its northeast extent, and the building of a permanent, heavy-duty road has begun toward the summit of Cadillac Mountain.

More significant changes to the park are evident on the USGS's next edition of this map, the edition of 1931, entitled *Topographic map, Acadia National Park, Hancock County, Maine.* First of all, there is the obvious change in the name of the park by this time. Secondly, this is a larger map than the previous edition, encompassing the larger extent of the park, though still at the scale of 1:62,500 and still based on surveys of 1901 and 1902. This map reflects the passing of the first legislation related to the park since its founding in 1919. P.L. 70-667, signed into law on January 19, 1929, authorizes the receipt of gift additions to the park beyond the limits of Mount Desert Island, making possible the acceptance in that same year of the gift of Schoodic Peninsula, on the east side of Frenchman Bay (45 Stat. 1083). The donors were influential in the change of the park's name from Lafayette to Acadia National Park.[65]

The 1931 map takes in the whole area from Blue Hill Bay, west of Mount Desert Island, to the east across the island and Frenchman Bay to include Schoodic Peninsula. In addition to the new park land on the peninsula, the first park lands west of Somes Sound appear on the 1931 map. One of these tracts necessitated federal legislation...land transferred from the Navy at the site of a radio station at Seawall (P.L. 71-248, 46 Stat. 377, May 23, 1930). Other western additions show up at Flying Mountain, Acadia Mountain, and Beech Cliff, and include shore frontage on Somes Sound and Echo Lake. On the east side of Somes Sound, the 1931 map shows the addition to the park of the west shore of Eagle Lake, the first segment of the rock bound east coast of the island, and, contiguous with the coastal segment, "the Bowl and Beehive tract" that had been the first major gift of land to the Hancock County Trustees of Public Reservations in 1908;[66] and at long last, the 1931 map is the first USGS map to include the summit of Cadillac Mountain within the boundaries of the park.

Later USGS maps of the park not only reveal changes that occurred in the park boundaries and the development of the park area, but also show changes in USGS map presentation including increases and decreases in scale, use of insets to show park areas beyond Mount Desert Island, use of green overprinting to designate wooded areas, and use of shading as a supplement to contours and spot heights to show relief.

The USGS *Topographic map, Acadia National Park and Vicinity, Hancock County, Maine*, edition of 1942, at the larger scale of 1:31,680, shows that extensive additions to park acreage were made between the early 1930s and early 1940s. The territory west of Somes Sound became a checkerboard of parkland. East of Somes Sound, substantial gifts were helping the park area to spread in every direction from the mountainous core to the Atlantic Ocean, Frenchman Bay and Somes Sound shores.

A large Rockefeller gift to the park in the southeastern quadrant of Mount Desert Island included additional spectacular ocean frontage. In connection with this gift, Rockefeller proposed to extend a scenic drive for automobile travel along the coast, taking in a stretch that was, at the time, Navy land.[67] P.L. 74-323, which became law on August 24, 1935, authorized the exchange of land between the Navy and the Department of the Interior, so that approximately 12 acres at the Otter Cliffs radio station could be turned over to Acadia National Park (49 Stat. 795). The park surrendered to the Navy an alternative site for a radio station on Schoodic Peninsula.[68] This same law also authorized the building of a bridge across the inlet at Otter Cove, part of the Rockefeller plan for the route of the Ocean Drive.

The USGS map edition of 1942 shows the completed Ocean Drive, including the bridge across Otter Cove, as well as the completed road up Cadillac Mountain. It portrays Schoodic Peninsula in an inset, shows a park road circling the peninsula's perimeter, and indicates the new radio station site (though without labelling) on the southwestern lobe of the peninsula.

The USGS first issued a map of Acadia National Park as a whole with green overprinting showing wooded areas in 1952, the "edition of 1942, reprinted 1952." Interestingly, an event occurred during the 1942 to 1952 interval that had a great impact on the wooded areas of Mount Desert Island and the park, but it is not clearly reflected on the 1952 reprint. A fire that started near Hulls Cove on October 21, 1947, spread, for a time with near-hurricane-force winds, predominantly to the south and southeast, sideswiping the village of Bar Harbor and devastating the eastern part of the park.[69] Four to five years of new growth may have been enough to bring back the wooded character of the eastern side of the island to allow for significant green on that part of the map.

Additions to and subtractions from Acadia National Park acreage also occurred in the 1942 to 1952 period, and these are reflected on the 1952 reprint of *Topographic Map, Acadia National Park and Vicinity, Hancock County, Maine.* One hundred fifty-two acres of additional park land on Schoodic Peninsula, adjacent to the radio station, were transferred to the Navy as a result of P.L. 80-260 of July 30, 1947 (61 Stat. 519). Sand Beach and the promontory of Great Head on the Atlantic coast of Mount Desert Island; Thompson Island at Mount Desert Narrows (where the island comes closest to the mainland); the western half of Bar Island, connected to Bar Harbor with the feature that gives the harbor its name; and a small tract of Little Cranberry Island, at the site of the Islesford Historical Museum, are among the extremities added to the park by 1952 that show up as part of the park on the 1952 map. A significant portion of Isle au Haut, approximately 20 miles southwest of Mount Desert Island, became part of the park by 1952,[70] but does not show up on this map as an inset or in any other form.

The next map of Acadia National Park, edition of 1956, is expanded to the scale of 1:24,000. The park map for the first time includes an inset of Isle au Haut, as well as one of Schoodic Peninsula. Among the other additions to the park notable on this edition are Bald Porcupine Island in Frenchman Bay, Baker Island lighthouse and its site at the center of Baker Island, and many acres on Mount Desert Island extending the park to the north, from Eagle Lake to the Hulls Cove area. In 1960 the 1:24,000-scale park map was reprinted with the addition of shaded relief.

The next USGS park maps were 1971 editions that returned to the smaller scale of 1:62,500. These were printed in 1972 with contours and in 1974 with the addition of shaded relief. Another issue of the 1971 map, with "minor corrections made 1976," was printed in 1977 at the scale of 1:50,000.

In the early 1980s, the first 1:24,000-scale 7.5 minute topographic quadrangle mapping of the area was completed with the publication of provisional and, in some cases, final editions of quadrangles with the following names: *Bar Harbor*, *Winter Harbor*, *Schoodic Head*, and *Seal Harbor* for the area of the *Bar Harbor* fifteen-minute quadrangle; *Newbury Neck*, *Salsbury Cove*, *Southwest Harbor*, and *Bartlett Island* for the area of the *Mount Desert* fifteen-minute quadrangle; *Swans Island* and *Bass Harbor* for the northern half of the *Swans Island* fifteen-minute quadrangle; and *Isle au Haut West* and *Isle au Haut East* for the southern half of the *Deer Isle* fifteen-minute quadrange. Some of these maps are metric editions, i.e., show elevations in meters.

MAPS FOR EXPLORING THE PARK BY ROAD, TRAIL, AND SEA

As basic and official as USGS maps are for the park, their appearance on the scene did not bring an end to the need for other kinds of maps of the park and its vicinity. For introductions to the highlights of the area and for guides to navigating by road, by trail, and by sea, there have been more tourist maps, road maps, trail maps and nautical charts published since the founding of the national park.

By 1939 the National Park Service was publishing a guide map to Acadia National Park.[71] A simple, uncolored map, it served an essential function of pointing out the park's major attractions and facilities to visitors, and showing them how to find their way via the main roads. At first it was published within a booklet with descriptive text, and later (by 1950), it was published as a brochure, with text on the map's verso. These early guide maps vary; relief is usually shown by hachures to give an idea of the locations of mountains, and routes of trails and carriage roads are sometimes included. The National Park Service guide map underwent periodic design changes, and in the early 1980s, a new design brought the park service guide map to where it is now: in color, with shaded relief and spot heights, with carriage roads and major trails shown in addition to the motoring roads, and with full color illustrations accompanying the text on the verso.

The USGS maps of the park, as much as they show trails, contours and physical features, cannot serve all the functions of a trail or path map…naming trails, giving distances, describing trail markings, keeping up to date. Harold Peabody and Charles H. Grandgent, in their guide to *Walks on Mount Desert Island, Maine*, make this point. After describing the Bates, Rand, and Jaques maps, and writing that their path maps of Mount Desert Island "are most useful with this book," they note that

> maps are also issued by the U.S. Dep't of Commerce [Coast & Geodetic Survey] and by the Geological Survey of the U.S. Dep't of the Interior. They are useful in studying the geography of the Island and adjacent lands and waters, but, having no trails marked, they are not of much use in following these walks.[72]

The later editions of the Bates, Rand, and Jaques path map, 1932 and 1937, extend geographic coverage beyond the eastern part of Mount Desert Island with a map on verso; *Path map of the western part of Mount Desert Island, Maine* at the scale of ca. 1:13,600 shows paths and trails maintained by the Southwest Harbor Village Improvement Association, within a portion of the island west of Somes Sound.

Augustus D. Phillips, of Northeast Harbor, created "path and road" maps of eastern and western Mount Desert Island in 1972, with new editions into the 1980s, reflecting the legacy of Bates, Rand, and Jaques. "In collaboration with Acadia National Park authorities for Mount Desert Chamber of Commerce" he portrays roads and named trails, shows relief by contours and spot heights, and includes park limits, placenames and points of interest. The eastern and western parts of the island appear back-to-back on two sides of the map sheet.

A relationship between the Appalachian Mountain Club (AMC) and Acadia National Park began to develop in 1922 when club members went to the park to camp on the shores of Echo Lake.[73] In 1925, the AMC camp on Echo Lake became permanent;[74] and thus the relationship became long-term. Almost forty years later, the first *A.M.C. Maine Mountain Guide* came out, describing hiking trails all over Maine. Among the separate folded maps issued with the guide was *The Appalachian Mountain Club Map of Mount Desert Island.* After the first edition in 1961, updated editions of the guide, with maps, have been issued in several subsequent years. In 1969 an excerpt of the guide, the *A.M.C. Trail Guide to Mount Desert Island and Acadia National Park,* was published separately; the 1968 edition of the map accompanies this excerpt. Later editions of the Mount Desert Island guide, with the map, have come out in 1971, 1975, 1984, and 1988. The AMC trail maps have maintained a clear and consistent style over the years. At the scale of 1:63,360, with relief shown by brown contours, waters in blue, and roads, trails, and placenames in black, these maps locate trails and paths on Mount Desert Island, without regard to park boundaries.

A recent addition to the mapping of Mount Desert Island, in the realm of basic topographical, road and trail maps, is the product of DeLorme Publishing Company of Freeport, Maine, which has become DeLorme Mapping Company. The company started its mapping activity in Maine in the late 1970s with an atlas of the state followed by separate sheet maps of several areas known for outdoor recreation. Beginning in 1982, DeLorme published *Topographic Map, Mt. Desert Island & Acadia National Park, Hancock County, Maine.* The credits state "mapped and edited by the Geological Survey, 1934-1935, 1939, and revised 1956; revised and reprinted by DeLorme Publishing Company." Like the USGS map of 1956, with shaded relief added in 1960, this map has a scale of 1:24,000 and relief shading. It is the commercial answer to keeping in print a large-scale shaded relief map of a national park that the USGS has discontinued from publication. It lacks insets of Schoodic Peninsula and Isle au Haut. Nonetheless, it is a basic reference map for paths, trails, roads, physical features, and park facilities on Mount Desert Island.

The United States Coast and Geodetic Survey until 1970, the National Ocean Survey from 1970 to 1980, and currently the National Ocean Service have frequently updated and reissued nautical charts of Acadia National Park's coastal waters. The chart-making agency changed the chart coverage areas and chart numbers somewhat from early editions. More recently, in 1974, the National Ocean Survey changed the chart numbers but maintained a constant configuration of charts in the Mount Desert Island region.

PICTORIAL AND SOUVENIR MAPS

From the 1920s onward, many delightful pictorial maps of Mount Desert Island and Acadia National Park have been created by local artists and mapmakers. Both colored and uncolored, historical and nautical in flavor, they give visitors reminders of the cultural and natural associations they may have experienced in the area; for the artists/cartographers, they seem to be expressions of affection for the area.

In 1928 Joseph Sims published a limited, hand-colored edition of his *Historical Map of Mount Desert Island, Maine.* Focusing on the island but including some surroundings, the map treats the historical theme with separate paragraphs on the French and English heritage as well as brief notes in geographically appropriate spots. Aspects of the decorative compass rose, ships, and borders are reminiscent of antique maps. The characteristic rounded shapes of the hills and mountains are portrayed pictorially; the mountains and other places are named, and elevations are indicated in many cases.

Another map reminiscent of antique predecessors is that of Luther Phillips, published in 1932, covering Mount Desert Island, Schoodic Peninsula, and some surrounding islands and mainland. Its historical notes are brief and few, but it is bordered by Champlain's words describing his discovery of Mount Desert Island, and within an inset is a redrawing of Champlain's 1607 map. Another inset shows a rendering of an early map of the new world, surrounded by figures of Native Americans such as those that might be part of a seventeenth century cartouche. Ships, compass rose, sea creatures, other animals, and trees reinforce the early map image.

In her 1933 *Map of the Mount Desert Region Showing Some Events from Viking to Aviator*, Margaret Fellows uses contemporary design to show historical events and current features of Mount Desert and nearby islands. Miniature figures in action, animals, plants, structures, names, and dates accompany the rounded mountain profiles common to many of these pictorial maps.

Augustus Phillips pays tribute to the men behind the founding and expansion of Acadia National Park on his 1966 map entitled *Sunrise on Acadia, this National Park Established in 1919*. As the title suggests, he has a different approach to the relief of the island; the sunrise puts a glow on the east sides of the mountains, while the west sides are shaded. Lighthouses have a glow of their own, and nautical illustrations, historical notes, and current places of interest are elements of the cartographic composition. Geographic coverage extends from Blue Hill Bay west of Mount Desert Island to the east side of Schoodic Peninsula; Isle au Haut is not included.

Jane Crosen of Brooklin, Maine, has recently created a series of Maine coastal maps with reference as well as souvenir value on postcards, posters, and t-shirts. Among the titles is *Approaches to Acadia and Mount Desert Island*. Covering Mount Desert Island, adjacent islands, and the mainland north to Ellsworth, this map includes placenames and major roads, and shows relief with hachures and places of interest with a system of symbols.

THEMATIC MAPS

For scientific, economic and environmental purposes, thematic maps of various sorts have been produced for Acadia National Park and vicinity; they document resources and can be an aid in the assessment of needs and the lay-out of plans for the future. One of Erwin Raisz's early contributions as a landform cartographer was his *View of Mount*

Desert Island, Maine, folded into his thesis on the geomorphology of the island, published in the *Annals of the New York Academy of Sciences* in 1929.[75] This is a map of the special sort made by Raisz: a view of landforms and a block diagram of the island's slopes, drainage, shoreline, and subsurface composition. Other maps within the paper show physiographic divisions, distribution of granite in the region, the trend of glacial striations on the island, pre-glacial drainage, and the shift of the divide.

The great fire of 1947 that damaged one-third of the forests of Acadia,[76] as well as personal property in and around Bar Harbor,[77] inspired maps in journalistic reports. Such was the map showing burned areas on Mount Desert Island, titled *Acadia National Park: 1947 Fire*, by Devereux Butcher, in *National Parks Magazine*.[78]

A. W. Küchler, scholar of vegetation mapping and vegetation mapper himself, mapped the vegetation of southeastern Mount Desert Island in 1953. At the scale of 1:25,000 and in color, he mapped the same area three times, according to three different classification systems. These maps serve as a basis for demonstrating a comparison between different vegetation map types, and the purposes to which the types may be suited, in Küchler's 1956 article, "Classification and Purpose in Vegetation Maps." He found southeastern Mount Desert Island to be an appropriate area to help him make his points because of its varied topography, its large number of vegetation types in a small area, its lack of urban development, and its good road network for accessibility during field work.[79]

Planning for preservation and wise use of resources has been a concern on Mount Desert Island at least since the beginning of this century.[80] An early planning document that includes maps was produced when the Bar Harbor Village Improvement Association called upon Charles W. Eliot, 2nd, a landscape architect, in the 1920s "for a survey of both the National Park and the island and for recommendations as to a future policy of development."[81] The report that resulted from Eliot's studies includes three maps: *Open Spaces - Mt. Desert Island*, showing parks, golf courses, reservations and the like; *Points of Interest - Mt. Desert - Wilderness Areas*, showing points of natural and historic interest and proposed wilderness areas; and *Com-*

munications - Mt. Desert - Villages, showing roads, trails, and areas of settlement and business activity.[82]

The National Park Service administration began to prepare master plans for the national parks also in the 1920s.[83] Maps serve a crucial function in helping to convey the rationale of a plan by providing a geographical framework for the consideration of available resources and future patterns of conservation and development. The master plan prepared in 1966 contains a variety of thematic maps, many at the scale of 1:80,000. The maps present existing and proposed land classification and use, utilities and communications, visitor use, and trails; resources of natural history, flora and fauna, and archaeology; and general development and resource management plans. A boundary plan touches on the issue of slating some lands for acquisition and some for deletion, and of seeking legislation to limit the size of the park.[84]

Maps of the 1972 master plan focus on the National Park Service objective of "establish[ing] a firm park management boundary within which are defined zones of varying degrees of acquisition and/or control."[85] Land ownership, land classification, proposals for zoning and acquisition of land, and general development plans for uses and facilities are mapped for the park at the scale of ca. 1:105,000, and for the region at the scale of ca. 1:300,000.

Because Congress had never authorized the Department of the Interior to purchase land for Acadia National Park, the park's growth had occurred through donations and exchange of lands only, i.e., somewhat haphazardly.[86] Federal legislation was necessary in order for the park service to achieve planning objectives such as those expressed in the 1972 master plan. Congress did eventually deal with the issue; it passed "An act relating to the establishment of a permanent boundary for that portion of Acadia National Park within the town of Isle au Haut" (P.L. 97-335, 96 Stat. 1627), October 15, 1982, and "An act to establish a permanent boundary for the Acadia National Park in the state of Maine . . ." (P.L. 99-420, 100 Stat. 955), September 25, 1986. These acts recognize permanent boundaries for the park as depicted on specified maps dated October 1981 and May 1986, respectively; they authorize the purchase of lands within these boundaries; they limit development on private holdings within the boundaries; and they authorize the acquisition of conservation easements on adjacent lands and islands. The maps are titled *Boundary map, Acadia National Park, Town of Isle au Haut, Maine* and *Acadia National Park Boundary Map.*

In the process of publishing thematic map series for the United States as a whole and for large regions of the country, some federal agencies have cartographically portrayed subjects of interest for the area of Acadia National Park. Among these series is the National Wetlands Inventory, and part of the inventory is *An Ecological Characterization of Coastal Maine*, by the United States Fish and Wildlife Service's Biological Services Program, Northeast Region.[87] It includes 1:24,000-scale 7.5 minute quadrangle maps that describe such conditions as underwater sediments, algae populations, and shrubs and forests along the coast. These maps were published in 1980 and cover Acadia National Park along with the rest of the Maine coast north and east of Cape Elizabeth.

Full color maps of the bedrock and surficial geology of Mount Desert Island have been produced by the Maine Geological Survey as part of its Bulletin 38. This 1988 publication, including text entitled *The Geology of Mount Desert Island: A Visitor's Guide to the Geology of Acadia National Park*, describes the rock types, the bedrock and glacial history of the area, and the geologic formations. It conveys this information graphically on the two maps at scale 1:50,000, and within the text, on uncolored maps of differing scales showing Isle au Haut and Schoodic Peninsula.[88]

A recent example of a commercially published thematic map is the 1989 *Wildlife Map of Acadia National Park: A Guide to Locating the Fauna of Acadia National Park* at the scale of 1:50,000. Published by a firm called American Nature Maps in Reston, Virginia, the map was designed by Phillip A. Schneider with data compiled by Robert P. Russell. The map points out areas appropriate for visitors to sight birds, mammals, reptiles, amphibians, and marine species. The base is a shaded relief map, including Mount Desert Island, Isle au Haut, and Schoodic Peninsula.

New mapping technologies, from aerial photography and satellite imagery to digitized maps and geographic information systems, are making possible the creation of greater varieties of thematic maps and base maps with customized parameters in Acadia National Park and all over the country. Digital data provided by the USGS and other government agencies allow planners and government, commercial and academic mappers to tailor maps to their needs. Acadia National Park personnel are already building data bases of the park's natural and cultural characteristics into geographic information systems for multiple future applications.

As the National Park Service and Acadia National Park planners develop a general management plan and grapple with policies to balance preservation and use in one of the smallest but most heavily visited national parks in the nation,[89] maps can serve as tools for planning to prevent overuse. Paradoxically, maps have been and continue to be a key to the use of Acadia National Park.

Park visitors today may or may not know about the projects going on behind the scenes to preserve this unique spot for them; they may or may not know the park's history. If they find their way to the Acadia National Park visitor center, they are exposed to a grandly scaled relief model of the park, built in the tradition of Robert Abbe; it orients them to the island's physiography. They may pick up a National Park Service guide map to tour the park by car. They may purchase a nautical chart, topographic quadrangle, or trail map if they have time to explore the park by slower means of locomotion. And when they leave the area, they may find a souvenir map...a poster, a pictorial map, a postcard map...to remind them of where they have been. Each of these types of maps has a history within the history of the Acadia National Park region and a lively present in the use and preservation of the park.

NOTES

1. Franklin K. Lane, Letter to Stephen T. Mather, May 13, 1918, in U.S. National Park Service, *Report of the Director of the National Park Service to the Secretary of the Interior for the Fiscal Year Ended June 30, 1918* (Washington: Government Printing Office, 1918), 273-276.
2. Ibid., 275.

3. Samuel Eliot Morison, *The Story of Mount Desert Island, Maine* (Boston: Little, Brown and Company, 1960), 7-8.

4. Samuel de Champlain, *The Works of Samuel de Champlain in Six Volumes*, H. P. Biggar, general editor; v. 1, translated and ed. by H. H. Langton and W. F. Ganong (Toronto: The Champlain Society, 1922), 1: 282-283.

5. Champlain's manuscript map is in the collections of the Library of Congress. See Walter W. Ristow and R. A. Skelton, *Nautical Charts on Vellum in the Library of Congress* (Washington: Library of Congress, 1977), 11-13. The Library of Congress issued a facsimile edition of the map in 1981 under the title *Map of the Northeast Coast of North America, 1607, drawn by Samuel de Champlain.*

6. Thomas Jefferys, *The American Atlas . . .* (London: R. Sayer and J. Bennett, 1776), pl. 7 [i.e. pl. 14 in NYPL copy].

7. Morison, *Story of Mount Desert Island*, 22-23.

8. Richard Walden Hale, Jr., *The Story of Bar Harbor: An Informal History Recording One Hundred and Fifty Years in the Life of a Community* (New York: Ives Washburn, 1949), 74-75.

9. Morison, *Story of Mount Desert Island,* 24, 72-73.

10. Ibid., 73.

11. Peter J. Guthorn, *British Maps of the American Revolution* (Monmouth Beach, N.J.: Philip Freneau Press, 1972), 27.

12. J. F. W. Des Barres, *The Atlantic Neptune* (London, 1780-1781). This description is based on the Astor copy of Des Barres' *Atlantic Neptune* in The New York Public Library.

13. Morison, *Story of Mount Desert Island*, 18, 25-26.

14. Ibid., 73.

15. Morison (*Story of Mount Desert Island*, 26) uses the spelling Salem Town, and dates his survey 1806. The spelling and date used here are taken from the map reproduction at the Northeast Harbor Library.

16. George B. Dorr, "Lafayette National Park," in U.S. National Park Service, *Report of the Director of the National Park Service to*

the Secretary of the Interior for the Fiscal Year Ended June 30, 1919 (Washington: GPO, 1919), 246.

17. George B. Dorr, *Acadia National Park: Its Origin and Background* (Bangor, Me., 1942), 28-29, 32-35. Also to be found in reprinted version, Dorr's *The Story of Acadia National Park: The Complete Memoir of the Man Who Made It All Possible, George B. Dorr, the "Father of Acadia"* with introduction by Frank J. Matter (Bar Harbor: Acadia Publishing Company, 1985), Book I.
18. Dorr, "Lafayette National Park," 246.
19. Walter W. Ristow, *American Maps and Mapmakers: Commercial Cartography in the Nineteenth Century* (Detroit: Wayne State University Press, 1985), 327-338.
20. See Sargent F. Collier, *Mt. Desert Island and Acadia National Park: An Informal History*, rev. and ed. by G. W. Helfrich (Camden, Me.: Down East Books, 1978), 38; Hale, *Story of Bar Harbor*, 148; and Cleveland Amory, *The Last Resorts* (New York: Harper & Brothers, 1952), 11, regarding use of the term "cottagers."
21. Peter J. Guthorn, *United States Coastal Charts, 1783-1861* (Exton, Pa.: Schiffer Publishing, 1984), 5.
22. Barbara B. McCorkle, "Where the Land Meets the Sea: America Charts its Coasts," *Meridian* 1(1989): 17.
23. Morison, *Story of Mount Desert Island*, 73.
24. U.S. Coast Survey, *Report of the Superintendent of the Coast Survey, Showing the Progress of the Survey during the Year 1856* (Washington, 1856), 29-30, 92; and idem, *Report of the Superintendent of the Coast Survey, Showing the Progress of the Work for the Year Ending November, 1849* ([Washington], 1849), 31st Cong., 1st Sess., Sen. Ex. Doc. No. 5, Sketch A.
25. U.S. Coast Survey, *Report of the Superintendent 1856*, 29.
26. Morison, *Story of Mount Desert Island*, 33-34, 37, 61; and Collier, *Mt. Desert Island*, 40.
27. U.S. Coast and Geodetic Survey, *Catalogue of charts and Other Publications, 1887* (Washington: GPO, 1887), 29.

28. U.S. Coast Survey, *Report of the Superintendent of the Coast Survey, Showing the Progress of the Work for the Year Ending November 13, 1848* ([Washington], 1848), 30th Cong., 2d Sess., Sen. Ex. Doc. No. 13, 2-3; and idem, *Report of the Superintendent 1849*, 3.

29. U.S. Coast and Geodetic Survey, *Catalogue of Charts, 1887*, 28-29, 31-32.

30. U.S. Coast and Geodetic Survey, *Catalogue of Charts, Coast Pilots, and Tide Tables* (Washington: GPO, 1916), 22-25.

31. See Amory, *Last Resorts*, 261-326; Lenard E. Brown, *Acadia National Park, Maine: History Basic Data* (Washington: National Park Service, Eastern Service Center, Office of History and Historic Architecture, 1971), 37-65; Collier, *Mt. Desert Island*, 36-74; Hale, *Story of Bar Harbor*, 126-189; Morison, *Story of Mount Desert Island*, 44-56; and George E. Street, *Mount Desert: A History*, new ed., rev. by Samuel A. Eliot (Boston: Houghton Mifflin Company, 1926), 269-296, for more on the development of tourism on Mount Desert Island, and on the development of Bar Harbor as a resort town, following the inspiration of Hudson River School artists.

32. Morison, *Story of Mount Desert Island*, 61, 46.

33. Clara Barnes Martin, *Mount Desert on the Coast of Maine* (Portland: B. Thurston and Co., Printers, 1867). See author's note on page following title page in subsequent editions.

34. Clara Barnes Martin, *Mount Desert on the Coast of Maine* (Portland: Loring, Short & Harmon, 1874).

35. The annual reports of the U.S. Coast Survey list private and government requests for Coast Survey information in an appendix under a heading such as "Information furnished from the Coast Survey office, by tracings from original sheets, &c., in reply to special calls . . ." See U.S. Coast Survey, *Report of the Superintendent of the United States Coast Survey, Showing the Progress of the Survey during the Year 1872* (Washington: GPO, 1875), 63-64. In the report for 1853, Superintendent Bache notes that "calls for information from the archives of the Coast Survey

have very much increased of late years . . . ; information from the archives . . . [is] at the disposal of those who desire it, at the mere cost of copying" (U.S. Coast Survey, *Report of the Superintendent of the Coast Survey Showing the Progress of the Survey during the Year 1853* (Washington, 1854), 11).

36. Hale, *Story of Bar Harbor,* 126-32; Morison, *Story of Mount Desert Island,* 49-50; Street, *Mount Desert: A History,* 276-96.

37. B. L. Robinson, "Edward Lothrop Rand," *Rhodora* 27(1925): 17-27.

38. Edward L. Rand and John H. Redfield, *Flora of Mount Desert Island, Maine: A Preliminary Catalogue of the Plants Growing on Mount Desert and the Adjacent Islands* (Cambridge: John Wilson and Son, 1894), title page.

39. Ibid., 28.

40. Ibid.

41. bid., 29.

42. Ibid., 30.

43. Ibid.

44. Ibid., 30-31.

45. Ibid., 32.

46. Charles T. Jackson, *First Report on the Geology of the State of Maine* (Augusta, 1837).

47. Nathaniel Southgate Shaler, "The Geology of the Island of Mount Desert, Maine," in *Eighth Annual Report of the United States Geological Survey to the Secretary of the Interior 1886-'87* ([Washington], 1889), part 2, pl. 64, 76.

48. George B. Dorr, *Acadia National Park: Its Growth and Development* (Bangor, Me., 1948), 5-24, or Dorr, *Story of Acadia National Park,* Book II, 5-24; and Hale, *Story of Bar Harbor,* 175-177.

49. The Bunker map is undated, but Library of Congress copyright records indicate 1911 registration.

50. Dorr, *Acadia National Park: Growth and Development,* 9.

51. Ibid., 14; and Hale, *Story of Bar Harbor,* 175-177.

52. Raymond B. Fosdick, *John D. Rockefeller, Jr.: A Portrait* (New York: Harper & Brothers, 1956), 303-304.

53. Rieley & Associates, *The Carriage Roads at Acadia National Park: Resource Study*, Preliminary Draft (Charlottesville, Va., 1988).

54. The monument was named for the person Dorr called "the founder of Acadia," Sieur de Monts, the Frenchman to whom the region, loosely referred to as Acadia, was granted by the French king; the man who brought Champlain to the area to explore and describe the coast. See Dorr, *Acadia National Park: Origin and Background*, 35-36; and Morison, *Story of Mount Desert Island*, 8.

55. Dorr, *Acadia National Park: Origin and Background*, 5-6. See also *The Hancock County Trustees of Public Reservations: An Historical Sketch and a Record of the Holdings of the Trustees* (Bar Harbor, 1939) for background on the trustees.

56. Dorr, *Story of Acadia National Park.*

57. Dorr, *Acadia National Park: Origin and Background*, 7-8.

58. *Hancock County Trustees: Historical Sketch*, 9.

59. Ibid., 16.

60. A partial list of the Sieur de Monts Publications appears in U.S. National Park Service, *Report of the Director of the National Park Service to the Secretary of the Interior for the Fiscal Year Ended June 30, 1919* (Washington: GPO, 1919), 351-352, and on the back cover or inside back cover of some of the publications themselves.

61. *The Coastal Setting, Rocks, and Woods of the Sieur de Monts National Monument*, Sieur de Monts Publications [4], (Washington: U.S. National Park Service, 1917), 4 and *The Sieur de Monts National Monument and the Wild Gardens of Acadia: A Reprint from the Journal of the International Garden Club*, Sieur de Monts Publications 22 (Washington: U.S. National Park Service, [1918?]), 18.

62. "Dr. Robert Abbe, Surgeon, Dies at 77," *New York Times*, 8 March 1928, p. 25, col. 5.

63. "Building of Arts" is barely discernible in the inscription of the Abbe relief model photographed for the Sieur de Monts Publica-

tions. The Building of Arts was a casualty of the Bar Harbor fire of 1947 (Hale, *Story of Bar Harbor*, 208-209).

64. Dorr, "Lafayette National Park," 249.

65. Dorr, *Acadia National Park: Growth and Development*, 25-31. In addition to Dorr's discussion of the park's name change, see also the official word on this subject in U.S. National Park Service, *Report of the Director of the National Park Service to the Secretary of the Interior for the Fiscal Year ended June 30, 1929, and the Travel Season, 1929* (Washington: GPO, 1929), 4.

66. Dorr, *Acadia National Park: Origin and Background*, 6.

67. Dorr, *Acadia National Park: Growth and Development*, 32-39.

68. Ibid., 38.

69. Hale, *Story of Bar Harbor*, 216-237.

70. John W. Webber, "New England Mountains: Isle au Haut," *Appalachia* 28 (n.s. 16) (Dec. 1950): 251.

71. American Geographical Society, Map Department, *Index to Maps in Books and Periodicals*, 10 vols. (Boston: G.K. Hall & Co., 1968), 1:8.

72. Harold Peabody and Charles H. Grandgent, *Walks on Mount Desert Island, Maine* (Boston, 1928), 3-4.

73. George B. Dorr, "Lafayette National Park," in U.S. National Park Service, *Report of the Director of the National Park Service to the Secretary of the Interior for the Fiscal Year Ended June 30, 1922 and the Travel Season 1922* (Washington: GPO, 1922), 152.

74. Z. Carleton Staples, "Echo Lake Camp," *Appalachia* 16 (1926): 367.

75. Erwin J. Raisz, "The Scenery of Mt. Desert Island: Its Origin and Development," *Annals of the New York Academy of Sciences* 31 (1929): 121-186.

76. L. F. Cook, "The 1947 Forest Fire Record: One Third of Acadia Burned," *National Parks Magazine* 22 (1948): 22.

77. Hale, *Story of Bar Harbor*, 216-237.

78. Cook, "1947 Forest Fire Record," 21.
79. A. W. Küchler, "Classification and Purpose in Vegetation Maps (with Three Separate Maps)," *The Geographical Review* 46 (1956): 157-58.
80. See Charles W. Eliot, *The Right Development of Mount Desert* ([Boston?], 1904).
81. Bar Harbor Village Improvement Association, *Report of the Thirty-Sixth Year of the Bar Harbor Village Improvement Association, Ending September First, 1927* ([Bar Harbor: The Association, 1927?]), 6.
82. Charles W. Eliot, 2nd, *The Future of Mount Desert Island: A Report to the Plan Committee, Bar Harbor Village Improvement Association* ([Bar Harbor]: Bar Harbor Village Improvement Association, 1928).
83. William C. Everhart, *The National Park Service* (New York: Praeger Publishers, 1972), 69, 85.
84. U.S. National Park Service, *The Master Plan for Acadia National Park* ([Washington?]: The Service, [1966]), [pl. 1].
85. U.S. National Park Service, *Master Plan, Acadia National Park, Maine* ([Washington?]: The Service, [1972?]), 3.
86. Ibid.
87. U.S. Fish and Wildlife Service, Biological Services Program, Northeast Region, *Ecological Characterization of Coastal Maine, Atlas* (Newton Corner, Mass.: The Service, 1980).
88. Richard A. Gilman et al., *The Geology of Mount Desert Island: A Visitor's Guide to the Geology of Acadia National Park*, Maine Geological Survey Bulletin 38 ([Augusta], 1988).
89. Richard Sassaman, "Piecing Together a Management Plan: Maine's Patchwork National Park," *Christian Science Monitor*, 22 November 1988, p. 21-22.

SELECTED CARTOBIBLIOGRAPHY

Abbe, Robert. *The Beautiful Island of Mount Desert*. . . [relief models]. Horizontal scale of some [ca. 1:40,000]. [Bar Harbor? 1916?-]. Models located at Acadia National Park Headquarters, Bar Harbor Historical Society, and Northeast Harbor Library. Also, illustrated in Sieur de Monts Publications nos. 4 and 22, as cited in note 61.

Acadia National Park, Mount Desert Island [relief model]. Horizontal scale [ca. 1:7,000]. [197-?]. Located at Acadia National Park Visitor Center, Hulls Cove.

Appalachian Mountain Club. *The Appalachian Mountain Club Map of Mount Desert Island* [map]. [ca. 1:63,360]. Boston: AMC, 1961. With title variation, also published 1968, 1971, 1976, 1984, 1988.

Bates, Waldron, Edward L. Rand, and Herbert Jaques. *Map of Mount Desert Island, Maine* [map]. 1:40,000. [Bar Harbor?: Bates, Rand, and Jaques?], 1896. Also published 1901? 1906, 1911, 1917, 1918? 1921, 1922. Published 1934 with title *Road Map, Mount Desert Island, Maine.*

Bates, Waldron, Edward L. Rand, and Herbert Jaques. *Path Map of the Eastern Part of Mount Desert Island, Maine* [map]. 1:25,000. [Bar Harbor?: Bates, Rand, and Jaques?], 1896. Published at 1:29,000 in 1901, 1903? 1906, 1909, 1911, 1913, 1917, 1921, 1922, 1926. Also published in 1932 and 1937 with map on verso, *Path Map of the Western Part of Mount Desert Island, Maine*, scale [ca. 1:13,600].

Bunker, W. G. *Map of Mount Desert Island, Maine* [map]. [ca. 1:47,000]. Bar Harbor: W. H. Sherman, [1911].

Colby, George N. *Atlas of Hancock County, Maine.* Ellsworth, Me.: S. F. Colby & Co., 1881.

Colby, George N., and J. H. Stuart. *Map of Mount Desert Island, Sorrento, Portions of Lamoine, Hancock and Frenchmans Bay and Adjacent Islands, Maine* [map]. [ca. 1:17,500]. [Ellsworth, Me.?]: Colby & Stuart, 1887.

DeLorme Publishing Company. *Topographic map, Mt. Desert Island & Acadia National Park, Hancock County, Maine* [map]. [ca. 1:24,000].

Freeport, Me.: DeLorme, 1982. Also published 1986. Published 1989 by DeLorme Mapping Company.

Des Barres, Joseph Frederick Walsh. *Charts of the Coast and Harbors of New England* . . . 1781. In: Des Barres, J. F. W. *The Atlantic Neptune.* London, 1780-1781. See especially chart of "Mount Desart Island" and surrounding waters and lands, including Schoodic Peninsula, at scale ca. 1:49,000, and chart of southern Penobscot Bay and Isle au Haut Bay, including "Isle Haut," at about the same scale. Both of these charts are dated 1776.

Eliot, Charles W., 2d. *Hancock County Trustees of Public Reservations Gifts of Land for Acadia National Park, Mt. Desert Island, Maine* [map]. Compiled 1935. [ca. 1:60,000]. Chester, Vermont: produced by The National Survey. In: *The Hancock County Trustees of Public Reservations: An Historical Sketch and a Record of the Holdings of the Trustees.* Bar Harbor, 1939.

Jones, John, and Barnabas Mason. "A Plan of the Islands Eastward Laying from Penobscot Bay & of the Granted Townships, with Their Distances and Bearings from Each Other and from the Continent which is Bordered or Shaded with Green, . . . February 13, 1765" [ms. map]. One mile or 80 ch[ains to the] inch. The Massachusetts Archives, Boston.

Peters, John, and James Peters. *A plan of the French grant on Mount Desert . . . Bluehill, 1807* [map]. 80 rods to an inch. [Northeast Harbor?]: Augustus D. Phillips, 1970. Original manuscript at Bar Harbor Historical Society.

Phillips, Augustus D. *Path and Road Map of the Eastern Part of Mount Desert Island, Maine; Path and Road Map of the Western Part of Mount Desert Island, Maine* [map]. [ca. 1:28,500; ca. 1:40,000]. [Northeast Harbor: Trails Committee], Mount Desert Chamber of Commerce, 1972. Also published 1975, 1986.

Porter, Seward. *Chart of the Coast of Maine* [hydrographic chart]. [ca. 1:63,360] and [ca. 1:180,000]. Nos. 4 and 5. Boston: T. Moore's Lithography, [1837].

Rand, Edward L. *Map of Mount Desert Island, Maine* [map]. 1:40,000. [Cambridge, Mass.?: E. L. Rand], 1893. Issued with: *Rand and John*

H. Redfield, Flora of Mount Desert Island, Maine: A Preliminary Catalogue of the Plants Growing on Mount Desert and the Adjacent Islands. Cambridge: John Wilson and Son, 1894.

Towne, Salem. ["Plan of the Western Half of Mount Desert Island, Maine, 1808"] [ms. map]. (80 rods to 1 inch). Original manuscript in Mount Desert Historical Society Museum, Somesville.

U.S. Coast and Geodetic Survey. *Coast and harbor charts* [hydrographic charts]. [Washington]: USC&GS. See, for example, *Blue Hill Bay and Western Part of Mount Desert Island, Maine*, chart no. 307, 1:40,000, 1885; *East Penobscot Bay, Maine*, chart no. 309, 1:40,000, 1882; *Frenchman's Bay and Eastern Part of Mount Desert Island, Maine*, chart no. 306, 1:40,000, 1885; *Mount Desert Island, Maine*, chart no. 292, 1:40,000, 1882.

U.S. Coast Survey. *Mt. Desert Island, Maine* [map]. 1:80,000. Chart no. 103[(2)]. [Washington]: USCS, 1875.

U.S. General Land Office. *Sieur de Monts National Monument, Mount Desert Island, Maine, Embracing the Island Summit and about Five Thousand Acres of Adjacent Lands* [map]. [ca. 1:70,000]. [Washington, 1917]. In: 39 Stat. 1785 (1916), facing page.

U.S. Geological Survey. *Acadia National Park and Vicinity, Hancock Co., Maine, 1971* [map]. 1:62,500. Washington: USGS, 1972. Also shaded relief ed., 1974.

U.S. Geological Survey. *Acadia National Park and Vicinity, Hancock Co., Maine, 1971, minor corrections made 1976* [map]. 1:50,000. Reston, Va.: USGS, 1977.

U.S. Geological Survey. *[15 Minute Series (Topographic)]* [maps]. 1:62,500. [Washington]: USGS. First editions of quadrangles covering the area are *Bar Harbor, Maine*, 1904; *Deer Isle, Maine*, 1904; *Mount Desert, Maine*, 1904; *Swan Island, Maine*, 1904.

U.S. Geological Survey. *7.5 Minute Series (Topographic)* [maps]. 1:24,000. Reston, Va.: USGS. *Bar Harbor, Maine*, 1982; *Bartlett Island, Maine*, 1981; *Bass Harbor, Maine*, provisional ed., 1983; *Isle au Haut East, Maine*, provisional ed., 1983; *Isle au Haut West, Maine*, provisional ed., 1983; *Newbury Neck, Maine*, 1981; *Salsbury Cove, Maine*, 1981; *Schoodic Head, Maine*, provisional ed., 1984; *Seal*

Harbor, Maine, 1983; *Southwest Harbor, Maine*, 1983; *Swans Island, Maine*, provisional ed., 1983; *Winter Harbor, Maine*, provisional ed., 1984.

U.S. Geological Survey. *Lafayette National Park* [map]. 1:62,500. [Washington]: USGS, 1922. Also 1922 ed. reprinted 1928.

U.S. Geological Survey. *Topographic map, Acadia National Park and Vicinity, Hancock County, Maine* [map]. 1:31,680. [Washington]: USGS, 1942. Also 1942 ed. reprinted 1952.

U.S. Geological Survey. *Topographic map, Acadia National Park and Vicinity, Hancock County, Maine, 1956* [map]. 1:24,000. [Washington]: USGS, 1959. Also rev. 1956, with shaded relief added 1960.

U.S. Geological Survey. *Topographic map, Acadia National Park, Hancock County, Maine* [map]. 1:62,500. [Washington]: USGS, 1931.

U.S. National Park Service. *Acadia National Park Boundary Map* [2 maps]. [ca. 1:110,000] and [ca. 1:45,000]. [Washington?]: USNPS, 1986. Blue line print.

Walling, H. F. *Topographical map of Hancock County, Maine* [map]. 1:70,000. New York: Lee and Marsh, 1860.

FOLLOWING THE RIVER THROUGH THE MOUNTAINS: MAPPING BIG BEND NATIONAL PARK

Robert Sidney Martin

Louisiana State University Libraries

Big Bend National Park lies in the great crook of the Rio Grande which gives the park its name. Although Big Bend is one of the larger units in the park system, occupying some 708,000 acres of Brewster County, it is only a small fraction of the vast part of west Texas known as the Transpecos.[1]

The Athabascan Indians who once called the Big Bend region home believed that after the Great Spirit had finished making the earth, after He had placed the stars in the sky, the birds in the air and the fish in the sea, He found He had a large supply of unused raw material left over. This He hurled to the earth in one great heap, creating the Big Bend.[2] The appearance of the land certainly lends credence to the creation myth.

The Big Bend country is a land of spectacular contrasts. It includes both lowland deserts and mountain forests. Travelers encounter debilitating heat and freezing cold, deadly drought and flash floods. Although there are many canyons, broad stream valleys, and great arroyos, there is very little permanent water. It is desolate and barren, forbidding and uncompromising.

The Rio Grande marks much of the boundary of the park, forming a great sweeping curve around the Chisos Mountains, which dominate the park's center. In its 230-mile course around the park the river has cut three magnificent canyons: Santa Elena Canyon, or the Grand Canyon of the Rio Grande, to the west; Mariscal Canyon, at the southern edge of the park where the river makes its big turn from the southeast to the north; and Boquillas Canyon to the east. These three chasms are the result of the river's erosion of three large upthrusts: Mesa de Anguila, Mariscal Mountain, and the Sierra del Carmen.

Within the park the elevations range from about 2,000 feet above sea level at the river's edge to the heights of Emory Peak at 7,825 feet. As a result of this variation, all of the biospheres from Chihuahuan Desert to Subalpine are supported in the park. More than four hundred species of birds have been documented within the park boundaries, and equally large samples of reptilian and mammalian wildlife are found. The variety of climates also supports a botanical system that is strange and wondrous, ranging from the desert cacti to forests of scrub oak, pine and the uncommon madroña.

The country's harsh terrain, scrubby vegetation, scarcity of water, and severe climate, coupled with its almost complete lack of any attractive mineral wealth, are the principal reasons why the Big Bend area remained truly one of the last frontiers. Long after most of the West had been explored and settled, the Big Bend remained a haven for bandits and outlaws. It was the heart of the area which the Spanish referred to as the *despoblado*...literally, uninhabited.[3] As this desolate piece of land remained inhospitable and uninhabited, it also remained largely unexplored and unmapped.

The maps depicting the area that is now Big Bend National Park follow quite well the general typology of the development of cartography of any region.[4] The maps of the earliest period were based on the reports of the first explorers, drawn from crude observations and lists of topographic features, and were therefore vaguely impressionistic, with few details. In the second period, the maps are based on actual, though crude, surveys: the details usually correspond with recognizable features and, although the topography is often distorted, the place names are frequently familiar. The maps of the final period are the result of the work of professional surveyors employing refined equipment and techniques. These maps provide accurate depictions of the land, assigning place names that are thoroughly familiar.

In applying this typology to the Big Bend, the maps of the earliest period are extremely vague about the course of the Rio Grande which forms its outline; the earliest even show the river flowing to the Gulf of California rather than the Gulf of Mexico. Later, as Spanish explorers reconnoitered the territory, these errors were corrected, but few details were revealed. The shape of the river that gives the park its name did not appear on any map until near the end of the eighteenth century. It was not until the beginning of the present century that the maps of the Big Bend itself finally entered into the final phase of true accuracy.

THE SPANISH EXPLORATIONS

The earliest penetration of the Big Bend region was an indirect result of the attempts of Spanish explorers to duplicate the success of Cortez in Mexico. One expedition, under Panfilo de Narvaez, seeking in Florida wealth to equal that of the Aztecs, met disaster on the Gulf coast of Texas in 1528. Among the survivors was Alvar Nuñez, Cabeza de Vaca, who with three others wandered for many years throughout Texas, trying to find their way back to Mexico. Cabeza de Vaca's route through the Transpecos is uncertain; he may have crossed through the present national park. He was eventually rescued by a Spanish party and returned to Mexico, where he reported on the customs and habits of the native peoples he had encountered.[5]

Cabeza de Vaca's reports stirred the interest of the Spanish authorities in the possibility of finding the treasures of another Mexico in the interior, but there were neither treasure nor slaves in the Big Bend, and Spanish efforts were focused elsewhere. In 1581, however, the Viceroy authorized an expedition into the Transpecos, in search not of riches but of souls. Fray Augustín Rodríguez led a group of Franciscan missionaries up the Rio Conchos to its confluence with the Rio Grande, which they named La Junta de los Rios. There they encountered a sedentary tribe of indigenous people they named the Jumanos and established a mission to minister to their spiritual needs. The expedition continued up the Rio Grande to New Mexico.

Rodríguez had established not only a mission on the fringes of the Big Bend; he had also established a reliable route from Mexico to New Mexico: up the Conchos to the Rio Grande at La Junta and then up the Rio Grande. This route was followed the very next year, in 1582, by an expedition under Antonio de Espejo. On his return Espejo proceeded down the Pecos to the vicinity of the Toyah Lakes and then struck out cross-country to La Junta, thus skirting the northern fringes of the Big Bend. The fringes of the Big Bend had been penetrated by Rodríguez and Espejo but it remained unexplored.[6]

The lack of knowledge concerning the region was prominently displayed on the maps of the period. The map of New Spain appearing in Girolamo Ruscelli's famous 1561 edition of Ptolemy...which was one of the earliest printed maps to highlight New Spain…does not even clearly identify the Rio Grande.[7] The one Spanish printed map of the period, appearing in Antonio de Herrera y Tordesillas's official history of the New World, provided no details of the interior at all.[8] The manuscript map made around 1600 by Enrico Martinez, depicting the explorations of Juan de Oñate in New Mexico, does clearly show the Rio Grande, which it labels the "Rio Bravo" and the "Rio del Norte," but there is no indication of any bend in the river between the confluence of the Rio Conchos and the Pecos (here labeled "Rio Salado").[9]

The route from Mexico to Santa Fe was soon shifted westward to El Paso, and Spanish activity in the region centered along the upper Rio Grande during most of the seventeenth century. The resulting lack of information about New Spain in general was demonstrated in the European maps of the period. The famous map of North America by Jan Janssen, which first appeared in his 1636 edition of Mercator's *Atlantis Novi*, is an excellent example: it clearly shows the Rio Grande flowing into the Gulf of California.[10] This error was repeated in the landmark map of New Mexico published by the renowned cartographer Nicolas Sanson d'Abbeville in 1656.[11]

In 1684, in response to a direct request from the Indians to reestablish the mission at La Junta, Captain Juan Domínguez de Mendoza led an expedition south from El Paso. Proceeding along the Rio Grande to La Junta, Domínguez continued overland toward the northeast in search of riches and treasure. He eventually penetrated as far as present Menard County before returning by the same route. He submitted a detailed report on his route and travels.[12]

It may be in part due to expeditions like that of Domínguez that the maps of the American Southwest began to show some improvement toward the end of the seventeenth century. Perhaps most typical was that of Vincenzo Coronelli, published in 1688, which in spite of its faults is a marked advance over its predecessors. The Rio Grande (here labeled the Rio Bravo) is shown correctly flowing to the Gulf of Mexico. In general form the river appears much like it does on Martinez's manuscript of 1600, but the Rio Conchos and Pecos do not appear, and there is no bend to the Rio Grande and no interior details in that vicinity.[13] Later printed maps improved somewhat on Coronelli's overall results but failed to advance his delineation of the Big Bend area. The landmark map of the area published by Guillaume Delisle in 1703 is almost identical to Coronelli's, adding only the Conchos River to the picture.[14]

Toward the end of the seventeenth century, provoked perhaps by harsh Spanish slave-trading along the Rio Grande, the hostile aboriginal peoples north of the river began increasingly to attack the Spanish settlements of northern Mexico. The focus of Spanish activities in the region therefore gradually shifted from treasure-hunting to defense, and punitive expeditions occasionally pursued Indians through the Big Bend region. One band of hostile natives, known as the Chisos Indians, was the focus of a series of expeditions in 1693 and 1694 which pursued them into the Big Bend area. Soon the Chisos were replaced as the dominant cause of trouble by the Mescalero Apache, who in turn were being pressured from the north by the Commanche.

Continuing defeats and wasteful expeditions eventually forced the Spanish to reconsider their approach to frontier defense. From 1724 to 1728, Brigadier Don Pedro de Rivera, under orders from the Viceroy, undertook an extensive inspection of all the presidios along the frontier of northern New Spain. Rivera did not penetrate the Big Bend, but his route circled almost completely around it. The geographical information from his tour was codified on a series of maps by Francisco Alvarez y Barriero, an engineer who had accompanied him on the expedition. Wheat states that Barriero's overall map of the territory, compiling into one chart the information from several others, was "a cartographic landmark of first importance" and claims that it was "by far the most accurate yet made of the area we know as the American Southwest." It is therefore all the more significant that Barriero showed no features and no bend in the Rio Grande between the mouths of the Conchos and the Pecos.[15]

In an effort to establish a barrier against the Indian depredations, Rivera decided to move the line of presidios on the frontier northward to the Rio Grande. Because this territory was completely unknown, Rivera dispatched seventy men under Captain José de Berroterán to find the best locations for the presidios. Following an extremely circuitous route, Berroterán eventually crossed the Big Bend from east to west on his return to La Junta. He was the first to cross the *despoblado,* but he found no useful presidio sites and demonstrated that La Junta was not approachable from the southeast by the river. In his report Berroterán claimed that the Apaches were able to raid the frontier with impunity because there were no presidios between El Paso and San Juan Bautista. He reported that the Indians retreated to this "unpopulated, long and wide gulf,"...the *despoblado*...which was uninhabited by settlers because of its "steep places, dry places, few waterholes, and great distances." He recommended establishing new presidios along the Rio Grande.[16]

The Spanish eventually adopted this suggestion. In 1747 Pedro de Rabago y Teran, Governor of Coahuila, led another expedition in search of a site for these presidios. It became the first Spanish expedition to enter what is now Big Bend National Park. Rabago y Teran marched west from Monclova, crossing into the Big Bend somewhere between Mariscal Canyon and Boquillas, probably near the present Rio Grande Village. He scouted into the Chisos but kept his main effort along the river. He reported sighting a double-shafted geological formation that could only be what is now known as Mule Ears Peak. He marched up Terlingua Creek, then west to the Rio Grand at Lajitas, and from there on to La Junta. He returned back to San Vicente via another route, north of the Chisos. He had explored much of the *despoblado* and had located waterholes and favorable routes. The only site the governor found for a presidio, however, was at La Junta, and one was duly established there in 1760.[17]

No maps appear to have resulted from Teran's expedition, but a short time later one of the most learned men in Mexico, Josef Antonio Alzate y Ramirez, published a map in Europe that purported to show all that was known by the Spanish in that area of the world. Alzate's map is a mass of confusions and distortions, showing for example the Pecos flowing into the Gulf of Mexico. Along the Rio Grande, below the mouth of the Conchos, there is the slightest hint of the great bend in the river that gives the region its name.[18]

In 1767 the Marques de Rubi, carrying out the instructions of the new Bourbon monarch Charles III, conducted an extensive inspection of the frontier of New Spain. The expedition toured the entire frontier from Texas to New Mexico, but bypassed the Big Bend. One result of this inspection was a magnificent map of New Spain by Nicholas de Lafora, Rubi's engineer, compiled in 1771. The map shows the province in great detail, but betrays the continuing Spanish ignorance of the Big Bend: the characteristic eponymous bend of the Rio Grande does not appear as the river continues its southeasterly progress almost without deviation. There are no details given in the interior of the area which is prominently labeled "Apaches Mescaleros."[19]

Rubi confirmed the Rio Grande as the defensive line and advised the creation of more presidios along the river to impede the depredation of the Apaches. In 1773 the Viceroy dispatched Hugo O'Conor to carry out Rubi's recommendations. O'Conor proceeded up the Rio Grande from San Juan Bautista establishing presidios in likely places. In the Big Bend area he established two presidios, one in the east, near a well-traveled Indian crossing of the Rio Grande, which he called San Vicente; and another in the west, about fifteen miles south of the river, which he called San Carlos. Both of these sites were south of the Rio Grande.[20]

In 1776 King Carlos reorganized his New World territories creating a single unit, the Provincias Internas, under which the entire northern frontier was administered. Teodoro de Croix was named commanding general of the provinces. After an extensive inspection, de Croix determined that, due to their isolation and poor communications, the presidios of San Vicente and San Carlos should be abandoned. Only La Junta remained in the Big Bend region.[21]

Indian depredations continued. Following a standard tactic, the Apaches were friendly to the soldiers garrisoned nearby but ventured far afield to pillage. Thus the governor of Coahuila, Juan de Ugalde, found it necessary to pursue marauding Indians into the *despoblado* which was not under his jurisdiction. Each year from 1779 through 1782, Ugalde undertook campaigns against the Apaches. These campaigns penetrated the Big Bend area and contributed substantially to the Spanish understanding of the geography of the region.[22]

Following the reorganization of the Internal Provinces, a new grand map of the region was constructed by a military engineer, Miguel Constansó, to display the boundaries of the various new departments. It depicts all of the vast territories of northern New Spain, from California to the Mississippi, in startling and accurate detail. The map was extremely accurate and served as the basis for several other important Spanish maps around the same time. One of these—known as the La Paz map because it is dedicated to the "Principe de La Paz"...incorporates much of the new information about the Big Bend region generated by Ugalde's punitive expeditions against the Apache. The general outline of the Big Bend area on the map is fairly accurate, and for the first time the sweeping curve of the river which gives the region its name is clearly portrayed. San Carlos and San Vicente are clearly shown south of the river, and the Chisos Mountains are properly located. The legend "Basilio" near the mountains is puzzling; perhaps it is somehow related to the feature in the center of the Chisos now known as the Basin.[23]

As Carl Wheat states, the Constansó map and the others based on it "mark the end of an epoch in Spanish discovery and exploration."[24] But a generation later another mapmaker added significantly to the record of exploration and discovery in the borderlands. In the first years of the new century Juan Pedro Walker, a young engineer of Anglo lineage, constructed a series of maps of the Provincias Internas, which stands as the epitome of Spanish geographical knowledge at the time of the Louisiana Purchase. One large map drafted by Walker in 1805 while he was assigned to the Presidio at Janos is particularly revealing. Along the Rio Grande, between the Conchos and the Pecos, significant cultural and topographical details are recorded, including for the first time the "Sra. de Chisos," in the middle of the great bend in the river. Walker's maps were never published, but they may have served as a basis for some important maps which came later.[25]

At the beginning of the nineteenth century, Spanish activity in the borderlands peaked and then began to wane as Spain began a slow world-wide withdrawal from its colonial territories. Throughout the period of Spanish hegemony in Mexico the Big Bend area had been essentially ignored. Spain quickly realized that it held no promise of riches or treasures but instead presented insurmountable obstacles at every turn. The established routes of communication between

Coahuila and New Mexico widely skirted the area, passing far to the south of the *despoblado*. Only when it became necessary to pursue and tame the nomadic tribes of Indians who sought refuge there did the Spanish attempt to establish any permanent presence in the Big Bend, and those attempts were doomed to failure. As the Spanish withdrew, the region remained essentially unknown.

AMERICAN EXPANSION

The Louisiana Purchase opened a new era in the exploration and mapping of the West but added little geographic knowledge about the Big Bend. Zebulon Montgomery Pike was dispatched in 1807 to explore the southwestern fringes of the new American territory. During his sojourn in Mexico, Pike acquired much new geographical information, and his report contained a map that was based on the best Spanish data he could acquire, including that of Juan Pedro Walker. The contours of the Big Bend on his map of 1810 are virtually identical to those on the La Paz map of thirty years before, and it includes no interior details.[26] The widely distributed map New Spain by Alexander von Humboldt, published about the same time as Pike's, was based on official sources including the maps of Constansó and was virtually identical to Pike's in the Big Bend area.[27]

The flood of Anglo immigrants into Texas during the 1820s and 1830s did little to add to the knowledge of the Big Bend region. The colonial ventures of Stephen F. Austin and his contemporaries were restricted to the area east of the 100th meridian. Austin's own important 1830 map extended only to the 102nd meridian.[28]

The Republic of Texas, established in 1836 after the successful Texas Revolution, claimed vast territories extending to the headwaters of the Rio Grande and beyond. Perhaps the most famous map of those claims is the one appearing in John Arrowsmith's *London Atlas* of 1841. The shape of the Big Bend on Arrowsmith's map remained impressionistic, and interior detail was almost completely lacking.[29] The map prepared in 1844 in the offices of the Topographical Bureau by William H. Emory is widely regarded as the epitome of the state of the knowledge of the region at the time. Emory relied heavily on

Arrowsmith's 1841 map for his own depiction of the Big Bend region, and the only significant new addition in the area is "The Trail of the Chihuahua Traders from Fulton in 1840."[30] This refers to an ambitious expedition under Dr. Henry Connelly, an American living in Chihuahua who hoped to initiate trade between Chihuahua and New Orleans which he believed would be more lucrative than the Santa Fe trade. The route of the expedition crossed the Transpecos, using a well-established trail which skirted the present park. The enterprise was not a commercial success and was not repeated.[31]

The war between the United States and Mexico, 1845-1848, resulted in numerous military expeditions in Texas and northern Mexico. Like similar ventures in earlier centuries, however, these forces found no reason to penetrate the great *despoblado*; all of the action took place far from the heart of the Big Bend country. The map published with the *Memoir* of Dr. Frederick Wislizenus is indicative of the geographical results of the war. A German physician, Wislizenus embarked on a private trip at his own expense, traveling from Independence, Missouri, to Chihuahua. His route through Santa Fe and down the Rio Grande demonstrates clearly the lengths travelers would go to avoid direct confrontation with the *despoblado*.[32]

The war concluded in February 1848 with the Treaty of Guadalupe-Hidalgo, the terms of which ceded vast new territories to the United States fulfilling the "manifest destiny" of the republic to reach from the Atlantic to the Pacific. Immediately authorities in Washington initiated an effort to reconnoiter this new land, attempting to locate the most expeditious routes linking the west with the east. Before the official explorers could arrive on the scene, however, a private venture financed by businessmen in San Antonio and led by Texas Ranger John Coffee Hays, set out toward El Paso in the spring of 1848, attempting to establish a trade route between the two cities. The men of the expedition endured many hardships, going many days without food and water. Their route took them up the Rio Grande from the mouth of the Pecos, over the Sierra del Carmen, and across to the Mexican side at San Vicente. They found the mountains on the right bank even more difficult than those on the American side but eventually reached San Carlos and continued on to Presidio del Rio.

They rested there for ten days and then threw in the towel, returning to San Antonio by the better-known northerly route skirting the Big Bend.[33]

No map resulted from Hays's exploits, but in his report on the expedition he pulled no punches in describing the territory. "The whole part of this country, from the mouth of the Devil's River up the Rio Grande, as far as San Carlos . . . is one constant succession of high broken mountains, destitute of timber and water," he wrote. It should not be surprising that he recommended the northern route by which he had returned as the better choice.[34]

The discovery of gold in California in 1848 imparted a new urgency to the need to establish a viable land route across the Transpecos. Capitalizing on the demand for information about the area, Robert Creuzbaur, the head draftsman of the Texas General Land Office, rushed into print with a guide book designed to encourage argonauts on the way to California to chose the overland route across Texas. The guide book included several maps, including "A Map to illustrate the most advantageous communication from the Gulf of Mexico and the Mississippi Valley to California and the Pacific Ocean." This work showed the Chihuahua trail...misplaced a bit too far south...but it gave absolutely nothing new about the Big Bend interior.[35]

In 1848 Lt. Col. Joseph E. Johnston was appointed to the post of chief topographical engineer in Texas with the responsibility of exploring, surveying, and mapping the vast uncharted territory. Johnston at first assigned his men to reconnoiter and map the routes already pioneered by a number of adventurers. These routes between San Antonio and El Paso all passed to the north of the Big Bend. Although these efforts added much to the information available about the Transpecos region, they did not penetrate the Big Bend proper. The map which Johnston compiled in 1850 and submitted with his reports, on which the various expeditions under his command are traced, clearly indicates the continuing lack of definition and detail on the interior of the great curve of the river.[36]

Johnston next turned his men to surveying the Rio Grande, in an effort to find out if the stream was navigable as far as the confluence with the Conchos as had been reported. It did not take the engineers long to disprove this theory. A team of engineers reconnoitering upstream from far down river at Ringgold Barracks was forced to turn back about eighty miles beyond the mouth of the Pecos. Another team floating downriver from El Paso abandoned the effort at Presidio declaring the canyons impassable by boat—and they had not yet seen Santa Elena, Mariscal, or Boquillas canyons. The results of these efforts were compiled on a single sketch. Although the general form of the area is evident on this map, there were still no details of the interior provided. The Big Bend remained unmapped.[37]

What Johnston's men had failed to do another party of men were forced to try only two years later. The Treaty of Guadalupe-Hidalgo had established a boundary between Mexico and the United States, stipulating that below El Paso the boundary was to be the center of the main channel of the Rio Grande. The treaty also required that the boundary be surveyed. The chief surveyor for the Boundary Commission, William H. Emory, selected M. T. W. Chandler, a young engineer from Philadelphia, for the task of surveying the river channel. On August 8, 1852, Emory ordered Chandler to commence the survey at Presidio and proceed downriver to the mouth of the Pecos, or until he met another party coming upriver. Emory warned Chandler that "the peculiar topographical features" of the region might force him to "deviate from the prescribed methods of operating." He gave Chandler a rough sketch of the country provided by a Mexican officer and assigned him a 35-man escort under Lt. Duff C. Green for protection from the Indians. Because Emory knew well the difficulty that lie ahead, he promised Lt. Green a promotion if the survey were completed successfully.[38]

The expedition encountered difficulties almost immediately, in the first canyons below Presidio in the Bofecillos Mountains. The surveying party on the river and the escort on the bank became separated, and all of the supplies and boats were lost in the rough water. After being re-supplied from Presidio the expedition pressed on, only to confront the difficulties of Santa Elena Canyon. Chandler dispatched a man into the canyon by foot, and when he returned it was agreed that the chasm was impassable. "It is impossible to keep along the edge of the stream in its course through the mountain," Chandler wrote in his report, "and just as impossible to navigate it. The falls and rapids which occur in quick succession, make the descent in boats entirely impractical." He decided to make as good a survey as possible without going through the canyon and instead detoured around it, climbing to the heights of the Mesa de Anguila from which he could see the river far below: "From the edge of the cañon the river may be seen far below, at a distance so great as to reduce it in appearance to a mere thread; and from this height the roar of the rapids is scarcely perceptible."[39]

After detouring around to the mouth of the canyon, the party proceeded downriver. At this point in his narrative Chandler commented on a particular mountain peak, visible for many miles, which was used again and again for sightlines and bearings. They soon found that this was the highest promontory in the Chisos Mountains and dubbed it "Mount Emory."[40]

Proceeding downriver, the expedition next encountered Mariscal Canyon. After losing one boat in the rapids, Chandler again concluded that "it was impossible to carry the line nearer the bed of the river than the summits of the adjoining hills" and detoured through the difficult mountains, eventually reaching the site of the abandoned Spanish presidio of San Vicente. He thought that the territory ahead was smooth and open country but he soon learned to the contrary, encountering the roughest terrain yet in the Sierra del Carmen range near Boquillas Canyon. Finding themselves in increasingly desperate shape and unable to carry on, Chandler and Green agreed to abandon the enterprise and found their way to Fort Duncan by way of a long detour through northern Mexico.

Chandler was awed by his experience and by the country he had seen. "No description can give an idea of the grandeur of the scenery through these mountains," he wrote in his report. "There is no verdure to soften the bare and rugged view; no overhanging trees or green bushes to vary the scene from one of perfect desolation."[41]

In spite of his inability to follow and map the river precisely, Chandler had accomplished much. His was the first scientific exploration and survey of the Big Bend. William H. Goetzmann has rightly concluded that the achievement of Chandler and his men "deserves to be ranked alongside those later, more famous, explorations of the canyons of the Colorado."[42] The results of his work were displayed on the map issued with the report of the Boundary Survey. Although the interior of the Big Bend remains devoid of detail, the course of the river is shown with great precision for the first time.[43]

One final and curious episode took place in the years immediately following the Boundary Survey. In 1855 Secretary of War Jefferson Davis authorized the importation of camels for use by the Army on the western frontier. It was hoped that these desert draft animals would prove more reliable in the arid conditions of the west than the traditional Army mule had. By 1857 a sizeable herd of the unwieldy beasts had been assembled at Fort Davis. In 1859 and 1860 successive expeditions under Lt. William H. Echols explored the Big Bend on camels in search of a site for a military post. These expeditions reconnoitered the areas to the east and the west of the Chisos Mountains, respectively. Although Echols' reports apparently included maps, they have not been located, and the data he reported apparently remained unpublished.[44]

The advent of war put an end to continued surveys and explorations in the West. Reconstruction further delayed additional reconnaissance in the Big Bend. The well-publicized exploits of Clarence King along the fortieth parallel in 1867-1868, of John Wesley Powell in the Grand Canyon in 1869, and of Ferdinand V. Hayden in the Yellowstone in the 1870s, however, contributed to a renewed interest in exploration and adventure. Texas Ranger Captain Charles L. Nevill attempted unsuccessfully to navigate the Santa Elena Canyon.[45] The establishment of Yellowstone as the first national park in 1872 kindled a spark in the Transpecos: Texans were hopeful that the vast unknown area called the Big Bend might prove equally spectacular. In 1883 the editor of the *El Paso Daily Times* appealed to other entrepreneurs to subsidize an expedition to the Big Bend, promising to expose the "sublime and majestic scenery" to the nation.[46] It was a good idea but far ahead of its time.

If great expeditions were not mounted to tame the Big Bend, still it began to surrender its wildness to the gradual process of civilization. The Army maintained a constant state of vigilance on the border, trying to hold down the effects of the continuing hostilities of the

Indians. Widespread ranches were established, small rural communities founded, and a level of commerce maintained. But still the interior of what was to be the Big Bend National Park remained largely unmapped.

SCIENTIFIC MAPPING

The next advance in the mapping of the park was not taken until near the end of the century, when the United States Geological Survey (USGS) began to devote some of its resources to the vast Transpecos region. Created in 1878, the survey gradually expanded its operations, mapping and investigating geological phenomena throughout the United States. By 1884 the survey extended its operations to Texas and found an almost complete dearth of information available. A special agent of the Census Bureau was retained to begin construction of a geological map.[47] By the following year there were three survey parties laboring in Texas, one devoted to triangulation and two to topographic work. Their efforts were concentrated in the area around Austin, where they succeeded in surveying and mapping 4,000 square miles.[48] Work continued in this manner for several years, with the survey parties gradually extending their efforts generally north and west from Austin. By 1892, 51,000 square miles had been surveyed and mapped in this fashion.[49]

In order to speed the work of the survey in Texas, in 1892 the number of working parties in the field was increased to five. Two of these, one for triangulation and the other topographic work, were designated to commence in El Paso and work east and south from that starting point.[50] In the years following, topographic sheets for El Paso, Marfa, Alpine, Valentine, Ft. Davis, and several fractional quadrangles along the Rio Grande were published.[51] For the first time accurate maps of much of the Transpecos were produced, but the work of the survey was then drawn elsewhere for a time. Although in the course of these mapping efforts the Big Bend itself must have received some cursory examination by the staff of the survey, their main work was still concentrated to the west and the east, and the park area remained unmapped.

This changed dramatically in 1899 when Robert T. Hill undertook the exploration of the Big Bend. Born in Tennessee, Hill was reared in Comanche, Texas, and early in life determined to be a geologist. With little preparatory education he worked his way through Cornell University, then the center of geological studies in America. During his junior year his interest in the geology of Texas and his achievement as a student came to the attention of Major John Wesley Powell, the director of the United States Geological Survey, who offered him a position with the organization. Hill joined the staff of the survey in 1885, and in the ensuing years gained both knowledge and stature in his studies of the Cretaceous system in Texas. Occasionally he was detailed to Texas for specific projects and, in 1893, he was designated chief of the survey's operations in Texas.[52]

Hill's investigations and resulting publications added greatly throughout the 1890s to the knowledge about the geography and geology of Texas. As the century drew to a close he concentrated his efforts on completing a monumental work on the physical geography of Texas, which would include a large map of the state, to be published as a folio in the survey's *Topographic Atlas of the United States.* In 1898 he undertook a general reconnaissance of Texas, "from Paris to El Paso," to gather information for the project. A draft of the map completed later that year by the survey's draftsmen, at a scale of twenty-five miles to the inch, "from the best available data, including a large amount of material furnished by Mr. Hill from personal observation."[53]

Hill had not yet completed his explorations, however. He had long been fascinated by the Big Bend and was particularly challenged by the fact that no scientific expedition had yet managed to force a passage of the canyons of the Rio Grande. "Since I first read the narratives in early childhood," he wrote in 1901, "they have excited my desire to try this passage. For years I have longed to launch a boat upon the river and to study the geologic story which the great walls of rock contain."[54]

Hill was undoubtedly spurred by the romance of the unknown and the thirst for new knowledge, eager for the opportunity to explore and map an unknown territory. But he was also motivated in part by the urge to be the first to explore the river by boat and the promise of fame that might result. In 1869 his friend and mentor, John Wesley Powell, had first explored the canyons of the Colorado River and had gained international renown as one of the great explorer scientists. Along the Rio Grande, in the Big Bend, Hill had the opportunity to accomplish a similar feat.

That Hill thought of himself in the same ilk as Powell can scarcely be doubted. The account of his adventure was published in *Century Illustrated Magazine*, which had also published Powell's account of the Grand Canyon expedition. Like Powell's, Hill's account was illustrated by artist Thomas Moran. In the opening paragraphs of his story, Hill made the following comparison:

> The wonders of the Colorado of the West were made known to the world through the dangerous trip of Major J. W. Powell in 1869, and are now brought within easy reach of the Pullman-car tourist. The cañons of the Rio Grande are the longest and least known; they have been and still are the least accessible to man, and have not hitherto been fully described.[55]

In 1899 Hill received authorization to explore the Big Bend country from Charles Walcott, who in 1894 had succeeded Powell as director of the Geological Survey. He enthusiastically set about making arrangements.[56] He recruited five other explorers and had three sturdy boats built and shipped to the expedition's jumping-off place at Presidio. The boats were packed with food, surveying gear, and photographic equipment. The only personal items allowed were those that would fit in the men's bedrolls. Fearing bandits and murderers as much as the river and the desert, plenty of guns and ammunition were also included.

The party set out on October 5, 1899. Soon they found themselves roaring through the canyons. Three exhausting days were required to portage the boats past the rock slide in Santa Elena Canyon but the men succeeded, becoming the first party to navigate the canyon. The expedition continued through Mariscal and Boquillas canyons, and ultimately on October 27 reached Langtry, Texas, where they found Judge Roy Bean, the Law West of the Pecos. "We had successfully navigated and mapped three hundred and fifty miles of a portion of one of America's greatest rivers which hitherto had been considered impassable," Hill wrote with pride in his account. "We had made a geologic section directly across the eastern sierra of the great American Cordilleras from the interior deserts to the coastal plain, procuring light upon some of our least-known country. . . "[57]

Hill's exploration did not gain him the fame of Powell and Hayden, but it was recognized as an important endeavor by the director of the Geological Survey. In his report Walcott noted the difficulty of the country through which Hill had passed, pointing out that "several parties, including those of the Boundary surveys, turned back after vain efforts to transverse it." He applauded Hill's efforts and noted that "the results of the expedition were of great interest, both geographically and geologically."[58]

The information that Hill obtained on the expedition was undoubtedly the basis for his "Reconnaissance Map of the Big Bend Country, Texas," which he compiled later that year.[59] The map, which covers virtually all of Brewster County at a scale of five miles to the inch, is unquestionably the first accurate map of what is now Big Bend National Park. All of the major topographical features of the park are evident, with the Chisos Mountains named in the center and the Rio Grande curving through the canyons around them. Persimmon Gap and Dog Canyon are featured as breaks in the Santiago Range (now known as the Dead Horse Mountains). The Rosillos Mountains are flanked by Bone Spring and Neville Spring. The Christmas Mountains and the Corazones Peaks rise to the northwest. Castelon Butte and Mule Ear peak rise southwest of the Chisos. The Terlingua Creek and Tornillo Creek drainages flow to the river on the west and east respectively. The Basin in the Chisos is clearly defined, although unnamed, as is the Mesa de Anguila. The three great canyons through which Hill passed are displayed in their rugged grandeur, though what is now known as Boquillas Canyon is here designated Carmen Canyon and Mariscal Mountain is named Big San Vincente Mountain. The village of Boquillas and the ruins of San Vicente are shown on the Mexican side of the River. Near the mouth of Terlingua Creek is a village named Dewese, now an abandoned site known as Terlingua Abaja. A road leads south from Marathon through Persimmon Gap, down Tornillo Creek to Boquillas. A branch forks west near Neville Spring, skirts the Chisos, and continues on to Terlingua

Creek, where it forks, with one branch continuing on to Terlingua and another heading north for Alpine.

This map was never published; it bears the stark notation by a Geological Survey bureaucrat "Do Not Engrave This" (fig. 4). The information it contained was given far wider distribution, however, through the much larger *Map of Texas and Parts of Adjoining Territories* which was issued as a part of the *Physical Geography of the Texas Region*, folio three of the *Topographic Atlas of the United States.*[60] Although drafted the year before, the map was not issued until April 1899, ample time to revise it to include the information from Hill's Big Bend explorations. The Big Bend portion of the map not only provides an accurate delineation of the river; it also supplies numerous interior details, although of course at a much reduced scale from the "Reconnaissance Map."

An impetus for further mapping in the Transpecos was soon provided by the search for mineral wealth. Reports of quicksilver in the Big Bend had circulated for years, but little evidence of the presence of the mineral in economic quantities was found until the 1890s. In 1896 the Marfa and Mariposa Mining Company was organized and established an operation near Terlingua, just northwest of the present national park. The mine was a great commercial success and stimulated additional ventures; by 1905 there were at least six mining companies in operation around Terlingua.[61] As William B. Phillips, the director of the University of Texas Mineral Survey, noted in 1901, the economic ventures resulted in pressure for maps:

> The importance of securing an accurate map of this part of Brewster country becomes at once apparent when it is considered that the deposits of quicksilver ores are extensive and valuable, and that a fair quality of coal can be obtained in that district.[62]

In December 1901, the University of Texas Mineral Survey entered into an agreement with the United States Geological Survey to map the Terlingua vicinity. The university provided approximately $2,500, and the USGS dispatched two of its experienced surveyors, Arthur Stiles and Fred McLaughlin, to carry out the survey. Stiles and McLaughlin went to work immediately, established initial vertical control points, and were in the field with plane table and theodolite

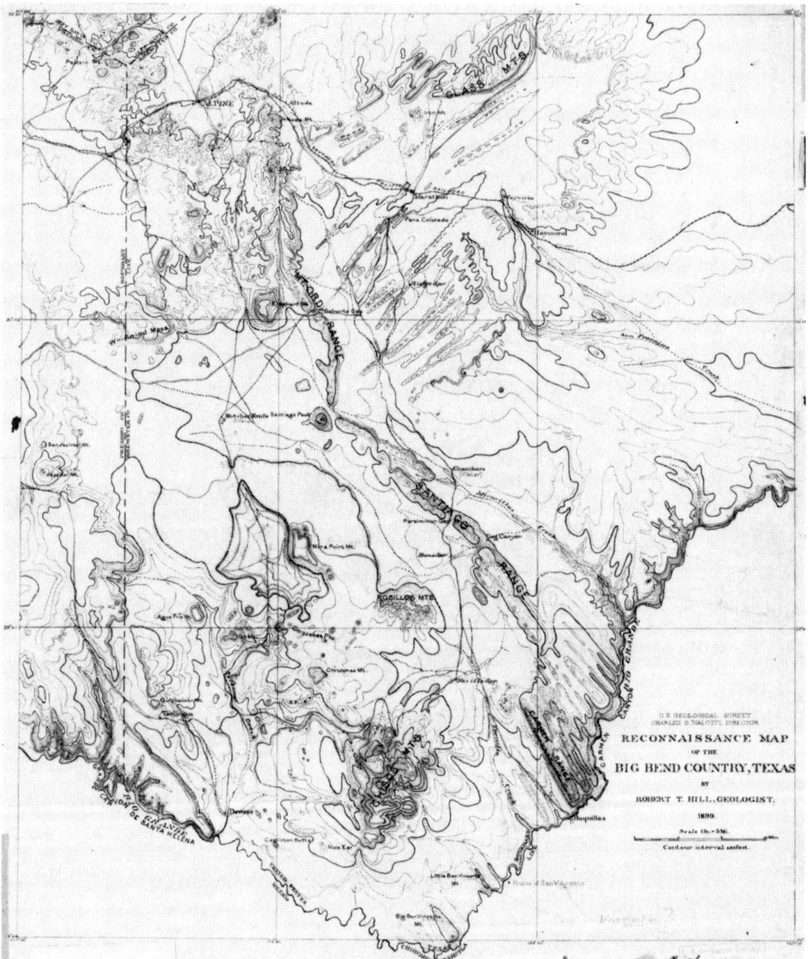

Fig. 4. Courtesy of National Archives and Records Administration.

by April 1902.[63] They rapidly completed their work, and 4,000 copies of the map of the Terlingua Special District were issued in November 1902. The map covered sixty-five square miles around Terlingua at a scale of 1:50,000 with a contour interval of twenty-five feet.[64]

The Terlingua Mining District lay entirely outside of the boundaries of the present national park, but immediately after completing the special mapping project, Stiles began work on the Terlingua and Chisos Mountains quadrangles in the regular thirty-minute USGS topographic series, thus bringing forward the topographic mapping that had halted short of the park area a decade before. In a short time he ran 123 miles of levels, established 34 permanent bench marks, and traversed 1,106 miles. The resulting maps, covering 898 square miles each, at a scale of 1:125,000 with a contour level of 100 feet, were published in 1903.[65] With the publication of these quadrangles the accurate mapping of all of what is now Big Bend National Park was finally accomplished.

New attention was focused on Big Bend when political strife in Mexico overflowed across the border in 1916. Bandits and revolutionaries staged a series of raids on border towns in Texas and New Mexico, and the United States Army under John Pershing was dispatched to the region to restore order. After the disturbances died down, soldiers who had been stationed in the area called attention to the rugged beauty of the land and its fitness for inclusion in the growing national park system. In the 1920s this idea grew into a groundswell of public opinion. In 1933 the Texas Legislature established Texas Canyons State Park in the Big Bend and appropriated funds to acquire lands. Not satisfied with this measure, boosters lobbied for the creation of a national park. The Park Service was persuaded to study the concept and, in 1935, issued a report recommending such a measure. Later that same year the required legislation was passed by Congress and signed by President Roosevelt. Big Bend National Park was a reality.

Even after the establishment of the park the 1903 USGS quadrangles by Stiles remained the standard maps of the area. These maps were not updated, revised, or replaced for almost seventy years, sound evidence of both their basic accuracy and the difficulty in mapping the rugged terrain of the Big Bend—as well as perhaps the lack of pressing need for more detailed maps. In 1969, in response to the growing demand from backpackers for topographic maps of the park, the Big Bend Natural History Association, in cooperation with the National Park Service, published an update of the two 1903 quadrangles on a single sheet. The only correction to the 1903 maps was the addition of the highways, roads, and foot trails in the park, as well as a detailed map of the Chisos Basin.[66]

Finally, in 1971, using modern mapping technologies, the USGS issued a new set of seven and one half minute quadrangles for the park. Twenty-seven quadrangles at a scale of 1:24,000 now provide detailed coverage of the natural and man-made features of the Big Bend.[67] At the same time a new, large scale topographic map covering the entire park at 1:100,000 was also published by the USGS.[68]

The story of the mapping of Big Bend National Park is not complete without some reference to geologic mapping. Although the geology of the region is unusual and interesting, like the topography it was left unexplored until the twentieth century. Robert T. Hill gave a general account of the area in a 1902 article on cinnabar deposits.[69] J. A. Udden presented an excellent summary of the basic stratigraphy and structure of the region as the results of a reconnaissance in 1907, but his report included no map.[70] As late as 1916, in a general survey of Texas geology, Udden reported that only exploratory mapping had been done in the Big Bend.[71] Brief mentions were made of the region in more general works thereafter, but it was not until general guidebooks for the park were created that the area received careful attention. The culmination of these efforts was the publication of a scholarly monograph on the Big Bend National Park in 1967 which included a large and detailed geological map. This map, by Ross Maxwell and others, established the basic geological framework which remains largely unchallenged today.[72]

The Big Bend was the heart of the *despoblado*. It remained essentially unmapped until the twentieth century because of its remote location, the rugged character of the terrain, and the apparent lack of mineral or other resources to attract interest. It was only on the eve of the twentieth century that the difficult canyons of the Rio

Grande were finally mastered and still later before the Geological Survey's field workers were able to overcome the obstacles it presented. When the Big Bend was finally carefully surveyed in 1903, the resulting maps were accurate enough to stand unrevised for more than half a century. The difficulty encountered by Arthur Stiles and his colleagues, mapping the region using the laborious techniques of the times, with compass and chain, plane table and theodolite, continue to command the respect of the cartographer, the historian, and the hiker alike.

NOTES

*This paper is based on a presentation given at the American Library Association Annual Conference, Dallas, Texas, June 1989.

1. The name "Big Bend" is sometimes loosely applied to the entire Transpecos region. Here it will be restricted to refer to the area within the actual boundaries of the national park which is the focus of this paper.
2. Ross A. Maxwell, *The Big Bend of the Rio Grande: A Guide to the Rocks, Landscape, Geologic History, and Settlers of the Area of Big Bend National Park,* Bureau of Economic Geology Guidebook 7 (Austin: University of Texas, 1968), 1.
3. An excellent general history of the Big Bend area is Ronnie C. Tyler, *The Big Bend: A History of the Last Texas Frontier* (Washington. D.C.: National Park Service, 1975).
4. See William P. Cummings' discussion in *The Southeast in Early Maps* (Chapel Hill: University of North Carolina Press, 1962), 1-4.
5. Carl O. Sauer, *Sixteenth Century North America: The Land and the People as Seen by the Europeans* (Berkeley: University of California Press, 1971), 117-22; Cleve Hallenbeck, *Alvar Nuñez Cabeza de Vaca: The Journey and Route of the First European to Cross the Continent of North America, 1534-1536* (Glendale, Calif.: Arthur H. Clark, 1940).

6. Carlos E. Castañeda, *Our Catholic Heritage in Texas, 1519-1936* (Austin: Von-Boeckmann-Jones, 1936-58), vol. 1: *The Mission Era: The Finding of Texas, 1519-1693* (1936), 157-194; Herbert E. Bolton, ed., *Spanish Exploration in the Southwest, 1542-1706* (New York: Scribner, 1916), 137-201.
7. Girolamo Ruscelli, *Nova Hispania Tabula Nova*, in Claudius Ptolemy, *La Geografia di Claudio Tolomeo Alessandrino, Nuovomente tradotta de Greco in Italiano da Girolamo Ruscelli . . .* (Venice, 1561). Reproduced in James C. Martin and Robert Sidney Martin, *Maps of Texas and the Southwest, 1513-1900* (Albuquerque: University of New Mexico Press, 1984), 68.
8. *Descripcion de las Yndias del Norte* in Antonio de Herrera y Tordesillas, *Descripcion de las Indias Occidentales* (Madrid, 1601). Reproduced in Martin and Martin, *Maps of Texas*, 76.
9. "Mapa del Nuevo Mexico: por el cosmográfo Enrico Martinez." The original is in the Archivo General de las Indias in Seville. See the discussion of the map in Carl I. Wheat, *Mapping the Transmississippi West, 1540-1861* (San Francisco: Institute for Historical Cartography, 1957-62), vol. 1: *The Spanish Entrada to the Louisiana Purchase, 1540-1804* (1957), 29-33. The map is reproduced by Wheat and is his number 34.
10. *Descriptio Americae Septentrionalis*, published in *Gerhardi Mercatoris Atlantis Novi Pars Tertia Italiam, Graeciam & maximas insulas Maris Mediterranei, nec non Asiam, Africam atque Americam continens Edito Ultima* (Amsterdam: Henrici Hondius, 1636). The map continued to appear in subsequent editions of the atlas for many years with some slight emendations. The bibliography of the atlas and the Mercator-Hondius-Janssonius succession are thoroughly discussed in Cornelius Koemann, *Atlantes Neerlandici: Bibliography of Terrestrial, Maritime, and Celestial Atlases and Pilot Books Published in the Netherlands up to 1880*. 5 vols. (Amsterdam: Theatrum Orbis Terrarum, 1967-1969), 2: 280-465.
11. *Le Nouveau Mexique, et La Floride: Tirée de diverses Cartes et Relations* published in *Cartès Generales des Toutes les Parties Du Monde* (Paris: Sanson, 1656). Reproduced in Martin and Martin, *Maps of Texas*, 82.

12. Bolton, *Spanish Explorations*, 320-43.

13. Vincenzo Maria Coronelli, *America Settentrionale colle nuove scoperte fin all'anno 1688* (Venice: Coronelli, 1688). Published in Coronelli's *Atlante Veneto, nel quale si contiene la descrittione geografica, storica, sacra, profana, e politica, degl'impreij, regni, provincie, e stati dell'universo* . . . (Venice: Coronelli, 1695). Reproduced in Martin and Martin *Maps of Texas*, 88.

14. Guillaume Delisle, *Carte du Mexique et de la Floride des terres Angloises et des Isles Antilles du course et environs de la Riviere de Mississipi* (Paris: Delisle, 1703). Reproduced in Martin and Martin, *Maps of Texas*, 51.

15. Wheat, *Mapping the Transmississippi West*, 1: 81; Francisco Alvarez Barriero, "Plano, corrografico é hidrografico, de las provincias de el nuevo mexico, sonora, ostimuri, culican, nueba vizacia [sic] najarit, nuevo rieno de Leon, nueva estramadura, õcoaguilla, y las nuevo reyno de Philipinas, provincia de los tejos todas de el minero de la nueva españa . . . ," original in the British Museum. The map is reproduced in Wheat and corresponds to his number 115.

16. Report of Captain Jose de Berroterán, quoted in James M. Daniel, "The Advance of the Spanish Frontier and the Despoblado" (Ph.D. diss., University of Texas, 1955), 186-187.

17. Castañeda, *Our Catholic Heritage*, vol. 3: *The Mission Era: The Missions at Work, 1731-1761* (1938), 211-223; Daniel, "Advance of the Spanish Frontier," 186-203.

18. Josef Antonio Alzate y Ramirez, *Nuevo mapa geografico de la America Septentrional, perteneciente al Virreynato de Mexico* . . . (Paris: Royal Academy of Sciences, 1768). Wheat number 149. Reproduced in Martin and Martin, *Contours of Discovery: Printed Maps Depicting the Texas and Southwestern Chapters in the Cartographic History of North america, 1513-1930* (Austin: Texas State Historical Association, 1982).

19. Nicholas de Lafora, "Mapa de la Frontera del Vireinato de Nueva Espana nuevamente contruido por el Yngeniero Ordinario D. Nicolas de Lafora sobre las observaciones de latitud que hizo.

Rumbos y distancias, que observo en su viage a aquellos paises, en el que manifesta el proyecto de so denfensa del Exmo. Sr. Marques de Rubi, y de dicho Yngeno. aprobado en la Junta de Guerra, y Hacienda, que se tuvo en esta Capital el dia 27 de Julio de 1771. Mexico a 30 Agosto de 1771." Madrid, Departmento de la Guerra. There is a full discussion of this map, and the several surviving copies of it, in Wheat, *Mapping the Transmississippi West*, 1: 87-89. The map cited corresponds to Wheat's 152 and is reproduced in Tyler, *Big Bend*, 39-41.

20. David M. Vigness, "Don Hugo O'Conor and New Spain's Northeastern Frontier, 1764-1776," *Journal of the West* 6 (Jan. 1967): 27-40. See also Sidney B. Brinckerhoff and Odie B. Faulk, eds., *Lancers for the King: A Study of the Frontier Military System of Northern New Spain, with a Translation of the Royal Regulations of 1772* (Phoenix: Arizona Historical Association, 1965).

21. Alfred B. Thomas, *Teodoro de Croix and the Northern Frontier of New Spain, 1776-1783* (Norman: University of Oklahoma Press, 1941), 22-58.

22. Daniel, "Advance of the Spanish Frontier," 263-67.

23. Miguel Constansó, "Carta ó mapa geográfico de una gran parte del Reino de Nueva España, comprendido entre los 19 y 42 grados de Latitud Septentrional y entre 249 y 289 grados de Longitud del meridiano de Tenerife, formado de orden del Excmo. Sr. Bo. Fr. Don Antonio maria Bucarely y ursa para indicar la division del Virreinato de Mexico y de las Provincias internas erigidas en Comandancia General en virtud de Reales Ordenes el año 1777." Archivo General de las Indias, Seville. The most well known of the subsequent maps based on the one by Constansó is "Mapa de las provincias internas de la America Septrional. Al Exmo Señor Principe de la Paz." American Antiquarian Society, Worcester, Mass. Wheat discusses these maps in *Mapping the Transmississippi West*, 1:121-24. The La Paz map is reproduced in Wheat (no. 195) and in Tyler, *Big Bend*, 46-47.

24. Wheat, *Mapping the Transmississippi West*, 1: 124.

25. Juan Pedro Walker, "Mapa Geografico Que comprehende de los terrenos de las Provincias de Texas, Coahuila, Nueva Biscaya y Nuevo Mexico; que de orden del Sõr. Conte Gral. de todas las Ynternas de Nva. España, Dn. Nemesio Salcedo y Salcedo, ha reconocido el Alferez de la Compañia presidial de Janos Dn. Juan Pedro Walker. Año de 1805." The original is in the Huntington Library; an almost identical map is also in the Barker Texas History Center at the University of Texas at Austin.

26. Zebulon Montgomery Pike, *A Map of the Internal Provinces of New Spain.* Published in Pike, *An Account of Expeditions to the Sources of the Mississippi, and through the Western Parts of Louisiana, to the Sources of the Arkansaw, Kans., La Platte, and Pierre Jaun, Rivers; performed by the order of the government of the United States during the years 1805, 1806, and 1807, And a tour through the interior parts of New Spain, when conducted through these provinces, by order of the captain-general, in the year 1807* (Philadelphia: C. & A. Conrad & Co., 1810). Reproduced in Martin and Martin, *Maps of Texas*, 110. Wheat devotes an entire chapter to Pike's explorations and reproduces the map (Wheat, *Mapping the Transmississippi West*, vol. 2: *From Lewis and Clark to Fremont, 1804-1845*).

27. Friedrich Alexander von Humboldt, *Carte Générale du Royaume de la Nouvelle Espagne. depuis le Parallele de 16· jusqu'ay Parallele de 38· (Latitud Nord) Dressée Sur des Observations Astronomiques et sur l'ensemble des Materiaux que existoient à Mexico commencement de l'année 1804.* Published in Humboldt, *Atlas Géographique et Physique du Royaume de la Nouvelle Espagne, fondé sur des Observations Astronomiques, des Mesures Trigonométriques et des Nivellemens Barométriques.* (Paris: F. Schoell, 1811). The London edition of the map is reproduced in Robert Sidney Martin and James C. Martin, *Contours of Discovery.*

28. Stephen Fuller Austin, *Map of Texas, with Parts of the Adjoining States.* (Philadelphia: H. S. Tanner, 1830). Reproduced in Martin and Martin, *Contours of Discovery.*

29. John Arrowsmith, *Map of Texas, compiled from Surveys recorded in the Land Office of Texas, and other Official Surveys*. (London: Arrowsmith, 1841). Reproduced in Martin and Martin, *Contours of Discovery*.

30. William H. Emory, *Map of Texas and the countries adjacent: Compiled in the Bureau of the Corps of Topographic Engineers from the best authorities, for the State Department. Under the direction of Colonel J. J. Abert. Chief of the Corps, by W. H. Emory. 1st Lieut. T.E.* (Washington, D.C.: War Department, 1844). Reproduced in Martin and Martin, *Contours of Discovery* and in Wheat, *Mapping the Transmississippi West*, no. 478, 2: 184.

31. Josiah Gregg, *Commerce of the Praries: or The Journal of a Santa Fe Trader, during eight expeditions across the Great Western Praries, and a residence of nearly nine years in Northern Mexico*. (New York: H. G. Langley, 1844), 162-165.

32. *Map of a Tour from Independence to Santa Fe, Chihuahua, Monterey and Matamoros by A. Wislizenus, in 1846 and 1847*. Published in Wislizenus, *Memoir of a Tour to northern Mexico, Connected with Col. Doniphan's expedition, in 1846 and 1847* (Washington: Tippin and Streeper, 1848). (30th Congress, 1st Session, Senate Miscellaneous Document 26)

33. A. B. Bender, "Opening Routes Across Texas, 1848-1850," *Southwestern Historical Quarterly* 37 (1933): 117-119; James Kimmin Greer, *Colonel Jack Hays: Texas Frontier Leader and California Builder* (New York: Dutton, 1952), 217-226.

34. *Democratic Telegraph and Texas Register* (Houston), December 28, 1848, January 25, 1849.

35. *Route from the Gulf of Mexico and the Lower Mississippi Valley to California and the Pacific Ocean, Illustrated by a General Map and sectional maps: with directions to travelers* (New York: Long and Brother, 1849). The map is reproduced in Wheat, *Mapping the Transmississippi West*, vol 3: *From the Mexican War to the Boundary Surveys, 1846-1854*, 68, (no. 597).

36. *Map—Reconnoissances of routes from San Antonio to El Paso Del Norte, &c., &c. By Brev. Lt. Col. J. E. Johnston, T. Engrs., Lt. W. F.*

Smith, Lt. F. T. Bryan, Lt. N. H. Michler, 1849. Including the Reconnoissance of Lt. W. H. C. Whiting, U.S. Engrs., 1849 (Washington, D.C., 1850). In *Reports of the Secretary of War, with Reconnaissances of Routes from San Antonio to El Paso by Brevet Lt. Col. J. E. Johnston . . .* (Washington, D.C., 1850). (31st Congress, 1st Session, Senate Executive Document 64.) For a complete discussion of the explorations and maps of the Topographical Engineers in Texas during this period see Robert Sidney Martin, "U.S. Army Mapping in Texas, 1848-1850," in Dennis Reinhartz and Charles C. Colley, eds., *The Mapping of the American Southwest* (College Station: Texas A & M Press, 1987), 37-56.

37. "Sketch of Reconnoissances of the Valley of the Rio Grande. April to September 1850. By Col. Johnstone [sic], T.E., Lt. W. F. Smith, T.E., Lt. Bryan, Lt. Michler, T.E., R. A. Howard, Esg., J. F. Minter, Esq.," Q 50, RG 77, Records of the Corps of Topographical Engineers, National Archives.

38. Ronnie C. Tyler, ed., "Exploring the Rio Grande: Lt. Duff C. Greens Report of 1852," *Arizona and the West* 10 (Spring 1968): 46-47.

39. M. T. W. Chandler to William H. Emory, December 1, 1852, in William H. Emory, *Report on the United States and Mexican Boundary Survey* (Washington, D.C.: Cornelius Wendell, 1857-1959) (34th Congress, 1st Session, House Executive Document 135), 1: 82.

40. Ibid., 83.

41. Ibid., 84.

42. William H. Goetzmann, *Army Exploration in the American West, 1803-1863* (New Haven: Yale University Press, 1959), 186.

43. Thomas Jekyll, *Map of the United States and their Territories between the Mississippi and the Pacific Ocean and part of Mexico, compiled from Surveys made under the order of William H. Emory. Major 1st Cavalry, U.S. Commissioner. And from the Maps of the Pacific Rail Road, General Land Office, and the coast Survey. Projected and Drawn under the supervision of Lt. N. Michler, Top. Engrs. by Thomas Jekyll, C.E.* (Washington, D.C.: S. Siebert, 1857-

58). Reproduced in Wheat, *Mapping the Transmississippi West*, 3:242 (no. 822).

44. "Diary of Lt. Hartz," in *Report of the Secretary of War* (Washington, D.C., 1858) (36th Congress, 1st Session, Senate Executive Document No. 2), 422-41; "Diary of Lt. Echols," *Report of the Secretary of War* (Washington, D.C., 1859) (36th Congress, 2nd Session, Senate Executive Document No. 1), 34-51. A good general account of the camel experiment is Frank Bishop Lammons, "Operation Camel: An Experiment in Animal Transportation in Texas, 1857-1860," *Southwestern Historical Quarterly* 61 (July 1967): 20-50.

45. *Austin Daily Statesman*, February 25, 1882.

46. *El Paso Daily Times*, June 18, 1883.

47. *Report of the Director of the Geological Survey* (Washington, D.C.: Government Printing Office, 1884), 40.

48. *Report of the Director of the Geological Survey* (Washington, D.C.: Government Printing Office, 1885), 12.

49. *Thirteenth Annual Report of the United States Geological Survey* (Washington, D.C.: Government Printing Office, 1892), 27.

50. Ibid., 80.

51. *Fifteenth Annual Report of the United States Geological Survey* (Washington, D.C.: Government Printing Office, 1894), 126; *Sixteenth Annual Report of the United States Geological Survey* (Washington, D.C.: Government Printing Office, 1895), 70; *Seventeenth Annual Report of the United States Geological Survey* (Washington, D.C.: Government Printing Office, 1896), 104.

52. Nancy Alexander, *Father of Texas Geology: Robert T. Hill* (Dallas: SMU Press, 1976), 3-115.

53. *Nineteenth Annual Report of the United States Geological Survey* (Washington, D.C.: Government Printing Office, 1898), 38-39.

54. Robert T. Hill, ""Notes on the Big Bend," (unpublished, completed 1901), quoted in Alexander, *Father of Texas Geology*, 122. See also Ronnie C. Tyler, "Robert T. Hill and the Big Bend," *American West* 10 (September 1973): 37.

55. Robert T. Hill, "Running the Cañons of the Rio Grande: A Chapter of Recent Exploration," *Century Illustrated Magazine* 61 (January 1901): 371. The ensuing account of the expedition is based entirely on this report.

56. Charles Walcott to Hill, May 19, 1899, cited in Alexander, *Father of Texas Geology,* 122.

57. Hill, "Running the Cañons," 387.

58. *Twenty-first Annual Report of the United States Geological Survey* (Washington, D.C.: Government Printing Office, 1900), 76.

59. Robert T. Hill, "Reconnaissance Map of the Big Bend Country, Texas . . . 1899" (RG 57, Records of the U.S. Geological Survey, National Archives).

60. Robert T. Hill, *Map of Texas and Parts of the Adjoining Territories,* in *Topographic Atlas of the United States: Physical Geography of the Texas Region* (Washington, D.C.: U.S. Geological Survey, 1900).

61. Kenneth Baxter Ragsdale, *Quicksilver: Terlingua and the Chisos Mining Company* (College Station: Texas A&M University Press, 1976), 13-21.

62. *Report of Progress for 1901. Sulphur, Oil and Quicksilver in Trans-Pecos Texas,* University of Texas Mineral survey Bulletin no. 2, issued as Bulletin of the University of Texas no. 9 (Austin: University of Texas, 1902), 4.

63. *Twenty-third Annual Report of the United States Geological Survey* (Washington, D.C.: Government Printing Office, 1902), 145.

64. *Twenty-fourth Annual Report of the United States Geological Survey* (Washington, D.C.: Government Printing Office, 1903), 151.

65. Ibid.; *Twenty-fifth Annual Report of the United States Geological Survey* (Washington, D.C.: Government Printing Office, 1904), 198.

66. Big Bend Natural History Association, *Big Bend National Park. Terlingua-Chisos Mountains, Texas.* (Big Bend, Texas: The Association, 1969).

67. The quadrangles are: Sauceda Ranch, Santana Mesa, The Solitario, Lajitas, Yellow Hill, Amarilla Mountain, Mesa de Anguila, Hen Egg Mountain, Terlingua, Castolon, Christmas Mountains, Tule Mountain, Cerro Castellan, Smoky Creek, Sombrero Peak, The Basin, Emory Peak, Reed Camp, Grapevine Hills, Panther Junction, Glenn Springs, Mariscal Mountain, McKinney Springs, Roy's Peak, San Vicente, Solis, Sue Peaks, and Boquillas Canyon.

68. *Big Bend National Park, Texas.* (Washington, D.C.: USGS, 1971).

69. Robert T. Hill, "The Cinnabar Deposits of the Big Bend Province of Texas," *Engineering Mining Journal* 47 (1902); 305-307.

70. J. A. Udden, "Sketch of the Geology of the Chisos Country, Brewster County, Texas," *University of Texas Bulletin* No. 93 (Austin: University of Texas, 1907).

71. J. A. Udden, C. L. Baker and Emil Bose, *Review of the Geology of Texas*, University of Texas Bulletin, no. 44 (Austin: University of Texas, 1916), 5.

72. Ross A. Maxwell, John T. Lonsdale, Roy T. Hazard, and John A. Wilson, *Geology of Big Bend National Park, Brewster County, Texas* (Austin: University of Texas, 1967).

SELECTED CARTOBIBLIOGRAPHY

Arrowsmith, John. *Map of Texas, compiled from Surveys recorded in the Land Office of Texas, and other Official Surveys* [map]. [1 inch = approx. 42 miles.] In: Arrowsmith, John. *London Atlas.* London: John Arrowsmith, 1841.

Austin, Stephen Fuller. *Map of Texas, with parts of the Adjoining States* [map]. [1 inch = approx. 24 miles] Philadelphia: H. S. Tanner, 1830.

Big Bend Natural History Association. *Big Bend National Park, Terlingua-Chisos Mountains, Texas* [map]. 1:130,000. Big Bend, Texas: Big Bend Natural History Association, 1969.

Canstansó, Miguel. "Carta ó mapa geográfico de una gran parte del Reino de Neuva España, compgrndido entre los 19 y 42 grados de

Latitud Septentrional y entre 249 y 289 grados de Longitud del meridiano de Tenerife, formado de orden del Excmo. Sr. Bo. Fr. Don Antonio maria Bucarely y ursa para indicar la division del Virreinato de Mexico y de las Provincias internas erigidas en Comandancia General en virtud de Reales Ordenes el año 1777." [ms. map]. Scale not given. Original in: Archivo General de las Indias, Seville.

Creuzbaur, Robert. No. 1. *A Map to illustrate the most Advantageous Communication from the Gulf of Mexico and the Mississippi Valley to California and the Pacific Ocean* [map]. Scale not given. New York: Mayer & Kroffs, 1849. In: Creuzbaur, Robert. *Route from the Gulf of Mexico and the Lower Mississipi Valey to California and the Pacific Ocean, Illustrated by a General Map and sectional maps: with directions to travelers.* New York: H. Long and Brother, 1849.

de LaFora, Nocholas. "Mapa de la Frontera del Vireinato de Nueva España nuevamente contruido por el Yngeniero Ordinario D. Nicolas de Lafora sobre las observaciones de latitud que hizo. Rumbos y distancias, que observo en su viage a aquellos paises, en el que manifesta el proyecto de so denfensa del Exmo. Sr. Marques de Rubi, y de dicho Yngeno. aprobado en la Junta de Guerra, y Hacienda, que se tuvo en esta Capital el dia 27 de Julio de 1771. Mexico a 30 Agosto de 1771." [ms. map] Scale not given. Original in: Departmento de la Guerra, Madrid.

Descripcion de las Yndias del Norte [map]. Scale not given. In: Tordesillas, Antonio de Herrera y, *Descripcion de las Indias Occidentales.* Madrid: 1601.

Emory, William H. *Map of Texas and the countries adjacent: compiled in the Bureau of the Corps of Topographic Engineers from the best authorities, for the State Department. Under the direction of Colonel J. J. Abert. Chief of the Corps, by W. H. Emory. 1st Lieut. T.E.* [map]. Scale not given. Washington, D.C.: War Department, 1844.

Hill, Robert T. *Map of Texas and Parts of the Adjoining Territories* [map]. 25 miles = 1 inch. Washington, D.C.: U.S. Geological Survey, 1900. In: Hill, Robert T. *Topographic Atlas of the United States: Physical Geography of the Texas Region.* Washington, D.C.: U.S. Geological Survey, 1900.

Hill, Robert T. "Reconnaissance Map of the Big Bend Country, Texas . . . 1899." [ms. map]. 1 inch = 5 miles. RG 57, Records of the U.S. Geological Survey, National Archives.

Humboldt, Friedrich Alexander, Freiherr von. *Carte Générale du Royaume de la Nouvelle Espagne. Dupuis le Parallele de 16º jusqu'ay Parallele de 38º (Latitud Nord) Dressée Sur des existoient à Mexico commencement de l'année 1804* [map]. Scale not given. Paris: 1809. In: Humboldt, Friedrich Alexander, Freiherr von. *Atlas Géographique et Physique du Royaume de la Nouvelle Espagne, fondé sur des Observations Astronomiques, des Mesures Trigonométriques et des Nivellemens Barométriques.* Paris: F. Schoell, 1811.

Jekyll, Thomas. *Map of the United States and their Territories between the Mississippi and the Pacific Ocean and part of Mexico, compiled from surveys made under the order of William H. Emory. Major 1st Cavalry, U.S. Commissioner. And from the Maps of the Pacific Rail Road, General Land Office, and the Coast Survey. Projected and Drawn under the supervision of Lt. N. Michler, Top. Engrs. by Thomas Jekyll, C.E.* [map]. 1:6,000,000. Washington, D.C.: S. Siebert, 1857-88. In: U.S. Boundary Survey. *Report of William H. Emory, Major First Cavalry and U.S. Commissioner.* Washington, D.C.: Cornelius Wendell, Printer, 1857. (U.S. House of Representatives, 34th congress, 1st Session, Ex. Doc. No. 135.)

Map of a Tour from Independence to Santa Fe, Chihuahua, Monterey and Matamoros by A. Wislizenus, in 1846 and 1847 [map]. 50 miles = 1 inch. Baltimore: E. Weber & Co., [1848]. In: *Mexico, Connected with Col. Doniphan's expedition, in 1846 and 1847.* (30th Congress, 1st Session, S. Misc. doc. No. 26.) Washington, D.C.: Tippin and Streeper, 1848.

"Mapa de las provincias internas de la America Septrional. Al Exmo Señor Principe de la Paz." [ms. map]. Scale not given. Original in: American Antiquarian Society, Worcester, Mass.

Pike, Zebulon Montgomery. *A Map of the Internal Provinces of New Spain* [map]. [1 inch = approx. 78 miles.] [Philadelphia: 1810] In: Pike, Zebulon Montgomery. *An Account of Expeditions to the sources of the Mississippi, and through the Western parts of Louisiana, to the sources of the Arkansaw, Kans., La Platte, and Pierre Jaun,*

Rivers; performed by the order of the Government of the United States during the years 1805, 1806, and 1807, And a Tour through the interior parts of New Spain, when conducted through these provinces, by order of the Captain-General, in the year 1807. Philadelphia: C. & A. Confrad & Co., 1810.

Ruscelli, Girolamo. *Nova Hispania Tabula Nova* [map]. Scale not given. In: Ptolemy, Claudius. *La Geografia di Claudio Tolomeo Alessandrino, Nuovomente tradotta de Greco in Italiano da Girolamo Ruscelli* . . . Venice: Ruscelli, 1561.

U.S. Geological Survey. *Big Bend National Park, Texas* [map]. 1:100,000. Washington, D.C.: U.S. Geological Survey, 1971.

U.S. Topographical Engineers. "Sketch of Reconnoissances of the Valley of the Rio Grande. April to September 1850. By Col. Johnstone [sic], T.E., Lt. W. F. Smith, T.E., Lt. Bryan, Lt. Michler, T.E., R. A. Howard, Esq., J. F. Minter, Esq." [ms. map]. Scale not given. Q 50, RG 77, Records of the Corps of Topographical Engineers, National Archives.

U.S. War Department. *Reconnoissances of Routes from San Antonio de Bexar to El Paso Del Norte, &c., &c., By Brev. Lt. Col. J. E. Johnston, T. Engrs., Lt. W. F. Smith, Lt. F. T. Bryan, Lt. N. H. Michler, 1849. Including the Reconnoissance of Lt. W. H. C. Whiting, U.S. Engrs., 1849*[map]. Scale not given. Philadelphia: P. S. Duval's Lith Steam Press, 1850. In: U.S. Senate. 31st Congress, 1st Session. *Reports of the Secretary of War, with Reconnaissances of Routes from San Antonio to El Paso by Brevet Lt. Col. J. E. Johnston* . . . (S. Ex. Doc. No. 64.) Washington, D.C.: the Union Office, 1850. (Serial Set no. 562).

AN UNDERGROUND ODYSSEY: THE DISCOVERY AND MAPPING OF CARLSBAD CAVERNS NATIONAL PARK

Heather F. Rex

Library
New Mexico State Highway and Transportation Department

Readers of history love a good exploration story, full of danger, discovery and unforgettable characters: stories of one-armed men like Captain John Wesley Powell hurtling down the Colorado River or stories of beautiful Indian princesses serving as wilderness guides to intrepid explorers from the other side of the continent. Such legends not only add to the world's body of geographic knowledge; the finely spun tale of exploration adds to the lore and the color of a place, enriching its cultural aura and giving that place qualities beyond the merely factual. Stories, whether fact or fiction, make it memorable.

The story of the exploration and mapping of Carlsbad Caverns National Park weaves fact into fiction so thoroughly that the fabric produced is a fine one indeed. What makes the tales from Carlsbad unique among mapping and exploration histories is the fact that they largely occurred underground, and maps play a far more significant role when the story unfolds in darkness.

Cave maps have multiple uses; the first use of maps was for exploration. Mapping a cave was the essential primary step in early cave studies because not only would the map show areas likely to lead to more passage, but it could possibly save the lives of the explorers. Although unexplored surface land may be approached without a map in hand such an omission would be foolhardy in a cave.

Cave exploration, by the very nature of the feat, is fraught with drama and danger. These inherent parts make it unlike any other kind of exploration. The personal danger is immense. Carried out in Stygian darkness with only small parts of the whole task ever revealed at any one time, cave exploration requires physical strength and emotional courage to endure the life-threatening and always present danger. Although explorers of a mountain must uncover its specific details, the mountain at least reveals its basic scope at the outset; the mysterious cave never does. Ronal Kerbo, former Cave Specialist at Carlsbad Caverns National Park, likens cave exploration to searching for the headwaters of the Nile: one knows something has to be up there (or down there in this case), but no one knows where. Cave explorers have no idea what they will find in their three-dimensional wilderness.

The differences in surveying and mapping above and below ground are as night and day...literally. In addition to being carried out in darkness, other obstacles to successful underground surveying are short sights, unsteady or steep ground for equipment and people, constricted passages, high humidity, and poor places for tripod placement, to name a few. Also, magnetic north must be checked frequently to identify anomalies that might affect the instruments. All equipment must be hand carried over sometimes great distances, lowered by ropes down steep drops and raised by ropes up daunting climbs. The environment can be uncooperative with the air containing dust or pulverized bat guano. It can be cool and damp with barometric winds reaching upwards of 65 miles per hour, as they can at Lechuguilla Cave.[1]

Cave surveying is simply a more difficult task than its surface counterpart, not only because of difficult conditions for humans but also because it has a more difficult subject to portray. The topographical surveyor only has to depict what exists on one plane (even though the topography exists in three dimensions), but a cave exists fully on three dimensions--passages have height as well as width and depth, and can occur one above the other. This third dimension must be depicted on the cave map. The cave map needs two basic drawings to be made, plan and profile views. A plan view will show the outline of the cave as seen from above, and the profile will show the cave as seen from the side (fig. 5).[2]

The surveyor has a considerable problem in presenting this information in a clear and concise way.[3] Understandably, the appearance of cave maps differs from surface maps because of their need to leave the two-dimensional and show both depth and height of passages. In their depiction of the third dimension, modern cave maps sometimes look like a strange hybrid cross of map and graph and engineering drawing with a y-axis used to impart height information. The symbology used on cave maps has its own uniqueness derived, once again, from the need to represent an additional dimension (fig. 6).

The mapping and exploring of ordinary caves can present heart-stopping challenges. Yet the seventy-five known caves in Carlsbad Caverns National Park are far from ordinary and their

A typical page out of a survey notebook showing station numbers, azimuth, distance, vertical angle, ceiling height, and passage width.

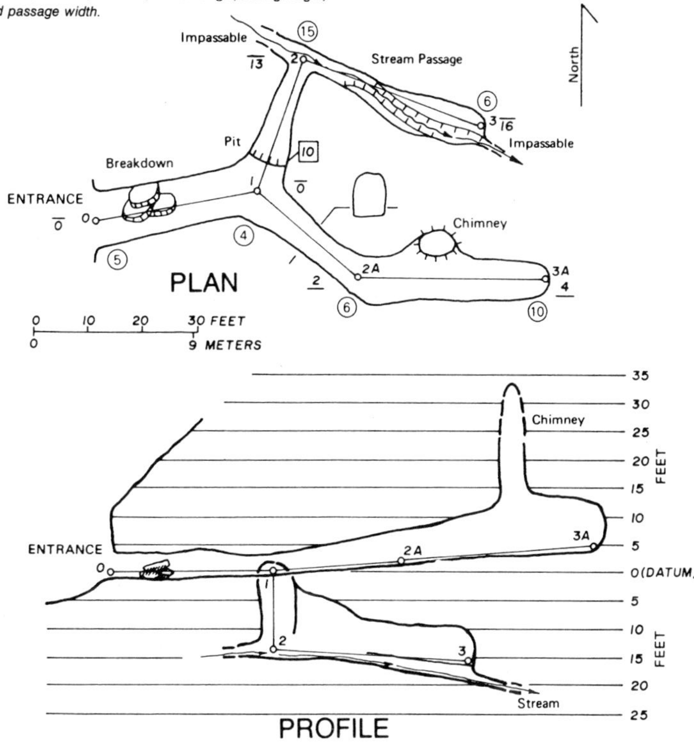

BAT CAVE BRUNTON COMPASS AND TAPE SURVEY, 7/18/77 BY P SLADE, J. SMITH, AND D BROWN			ALL AZ TO TRUE N DECL. CORRECTION 21° EAST		
STA.	AZ.	DIST.	VERT.	HT.	WD
0	80°	30'	0°	5	10
1	20°	25'	-30°	4	12
2	110°	36'	-5°	15	4
3	—	—	—	6	7
1-2A	130°	24'	+5°	4	12
2A	90°	35'	+3°	6	8
3A	—	—	—	10	6

Fig. 5.

exploration stories are likewise extraordinary. Perhaps the difficulty of mapping the caverns at Carlsbad is one reason relatively few early maps exist before the middle of this century.

Of the numerous caves in the park, Carlsbad Cavern was the first to be explored, is the best known, and is the one that lends its name to the park as a whole. Also referred to as the "main cave" and in early writings as "Bat Cave," Carlsbad Cavern's several names illustrate the interchangeability of the words "cave" and "cavern." Although it contains thousands of surface acres, the history of the exploration and mapping of Carlsbad Caverns National Park, is largely a history of subterranean adventures, a history that can be as murky as the caves themselves. Much early information about the main cave came from guides "who believed tourists appreciated hair-raising yarns more than hair-splitting distinctions. [They] did not allow dull fact to interfere with a good story."[4] Carlsbad Caverns' history is indeed a fine tale abounding with drama, peril, charlatans, facts, and legends.

NATURAL AND CULTURAL HISTORY OF THE PARK

Carlsbad Caverns National Park lies in southeastern New Mexico in the lower Pecos River Valley along the southeast facing exhumed reef escarpment of the Guadalupe Mountains, 22 miles southwest of the town of Carlsbad in Eddy County. With heights from 3,600 feet to 8,700 feet, the area is dominated by sharp limestone ridges and steep-walled canyons overlooking the gypsum Delaware Basin 1,000 feet below. The reef, called the Capitan Limestone, developed around the perimeter of the Delaware Basin when the basin was a warm and shallow Permian sea.

The caverns were produced by a not-fully-understood sequence of depositional, erosional, solutional, and uplift processes over the next 200 million years. Formation processes involved in the creation of the caves are still partly theoretical. Most of the caverns are limestone, as are 95 percent of the world's caves, but Carlsbad Caverns are geologically unique among the world's caves, a uniqueness that adds to exploration difficulties.

Usual cave formation involves the action of rain water, ground water, and carbonic acid on limestone, but at Carlsbad there is recent surprising evidence indicating these caves were formed through the action of sulfuric acid.[5] While most caves have long, tubelike passageways dissolved by underground springs, the caves in the Guadalupe Mountains consist of large, interconnected rooms of various and irregular profiles at several depths, spaces that sometimes lose their cave-like quality when ceilings soar to 300 feet. The main cave's Big Room has a floor that would hold fourteen football fields and a ceiling that is 22 stories high. In a more typical cave, it is easy to determine the points through which groundwater recharge takes place, thus indicating that cave's creation history. By contrast most caves at Carlsbad, long abandoned by the water that formed them, give few clues to their original water sources or flow patterns. Usual cave formation theories dealing with circulation of ground water and the carbonic acid reaction are insufficient to explain the unusual character of Carlsbad's cave complex.[6] Another result of their unusual geology and formation processes is that Carlsbad's caves are considered among the most spectacularly beautiful places on earth and well-deserving of their status as one of the wonders of the world.

Located at approximately 32°15' north latitude and 104°30' west longitude, the surface climate is semi-arid and continental with mild winters and sporadic rainfall resulting in the hardy, thorn-bearing vegetation typical of the Chihuahuan Desert. Drainage is mostly to the southeast into the Pecos River valley. Today the whole park encompasses 46,755 acres of rugged mountains and desert, with over 33,00 acres as designated wilderness. Teeming with life, a total of 740 species of plants, 273 species of birds, 59 species of mammals, and 44 species of reptiles have been identified.[7] The most famous of the mammals, residents of a portion of the main cave during the summer and numbering between 300,000 and 1,000,000, is the large migratory bat colony. Their hour-long twilight flight from the entrance, spiraling up and out, draws hundreds of thousands of visitors a year. The spectacle can be seen from miles away and looks from a distance like a huge billow of smoke.

Shown here are a few of the more commonly used cave map symbols.

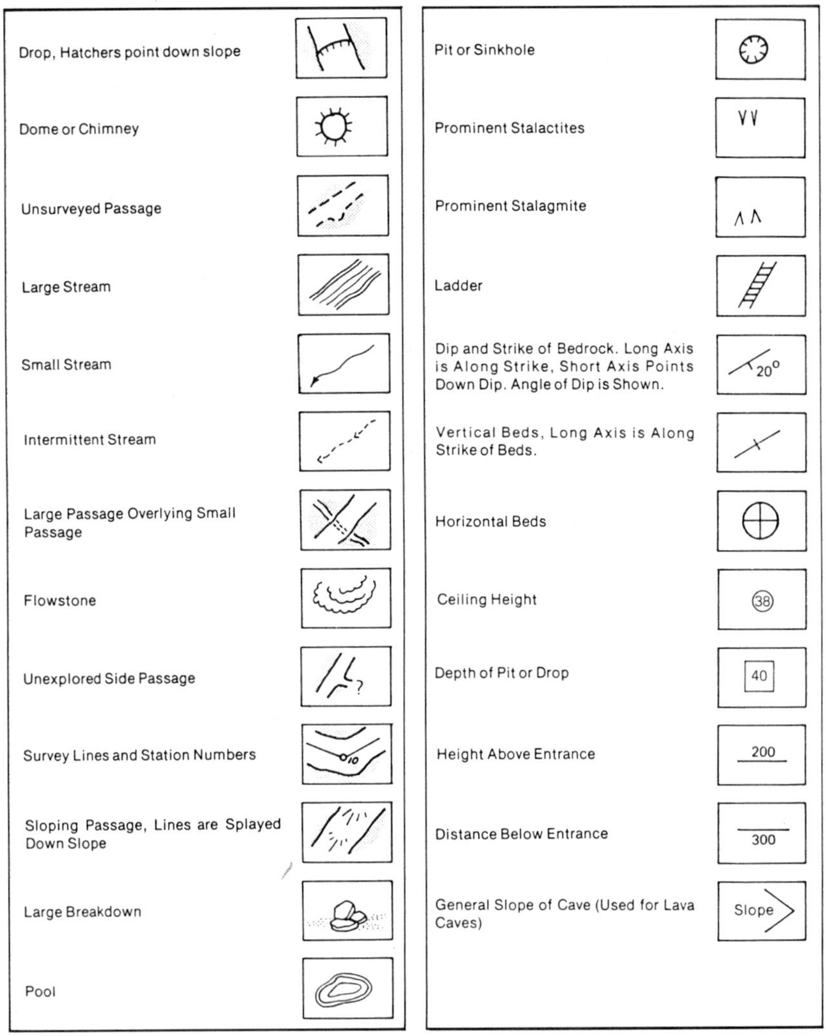

Fig. 6.

CULTURAL HISTORY OF THE CAVERNS

Earliest human occupation of the Guadalupe Mountains district probably occurred 12,000 years ago. Remains of human culture have been found in definite association with fossil bones of several extinct species, ground sloth, American horse, camel, caribou and muskox, as well as bison, antelope, wolf, and California condor.[8] Circular cooking pits, remains of a sandal in the entrance to the main cave, and wall paintings place a Basket Maker culture in the area of the caves approximately 1,500 years ago. However, there is no evidence that early man explored beyond the entrance. A human skeleton found 80 years ago in lower depths is presumed to be that of a prehistoric man who had fallen. Later Indians, those who lived around the caves between 600 and 800 years ago, received pottery and other objects from the Mogollon to the west, but there is no evidence that the Mogollon lived at Carlsbad, only of their cultural influence through the spread of ideas and trade objects.

During their exploration of New Spain in the sixteenth century, the Spanish reported Apaches living in the region. In all probability the Apaches, a hunting, nomadic people, began using the park area as camping sites in the fourteenth or fifteenth centuries. A legend of the Mescalero Apache Tribe is that of an ancient medicine man who desired to make "Big Medicine" by entering the cave. He was lost and never heard from again, but was seen from time to time wandering from the opening beating his drum. For many years on the date of his disappearance, the Apaches would come to the entrance and leave him food. Contemporary tribe members will still occasionally ask park personnel if they hear drums on the anniversary of the medicine man's exploit.[9]

The first historic record of the region was compiled by the conquistadors whose expeditions from Mexico City northward traversed the Pecos Valley. It is believed they did not reach the park area proper, although they passed nearby in both 1535 and 1583. The park area as well as a large part of the southwestern United States was under the Spanish flag from the time of the conquistadors in the sixteenth century until Mexico won its independence from Spain and took over the region in 1821. The Treaty of Guadalupe Hidalgo in 1848 which ended the Mexican American War granted much of today's Southwest to the United States. Congress created the territory of New Mexico in 1851, and the territory was admitted to the Union as the forty-seventh state in 1912.

Following New Mexico's designation as a territory, numerous United States military expeditions crossed the region and the relative calm of Spanish and Mexican rule came to an end. The first mention in print of any caves in southeastern New Mexico was by Captain John Pope in 1854 as he surveyed a route for the Pacific Railroad.

> Along our road today we discovered several caves. On examination they proved to be highly interesting, their interior being composed of gypsum, the roofs and sides highly polished and of purest white. They will repay the trouble (no easy matter) of entering them.[10]

The "highly interesting" caves referred to in Pope's journal lie in the gypsum beds outside the actual park and were not the even more astonishing caves found later within the park, but the reference is undoubtedly the first documented notice of the region's numerous caves.

The Butterfield Trail, route of the overland mail from St. Louis to San Francisco, passed just south of the Carlsbad Caverns area for approximately one year before the route was moved even farther south into Texas. Evidence of the exact location of the route through the Delaware Basin is sketchy because no stage station seems to have been established, but the Butterfield Trail probably came no closer to the park than thirty miles.

The first permanent Anglo settlements in southeastern New Mexico were established by 1855 at Fort Stanton and at Lincoln (famous for the death of Billy the Kid) in the Capitan Mountains to the northwest. It was not until after the Civil War that civilian settlers arrived in the lower Pecos Valley. Those settlers, ranchers and farmers, were associated with the vast land holdings of John H. Chisum around Roswell, New Mexico. Their first glimpse of the park area remains undocumented.

However, by the 1880s settlers must have entered the park because within a few years they requested a survey. Township 24 South, Range 25 East, was surveyed by Alfred H. Warren of the General Land Office from July 10 to 16, 1884. Although Section 31 of that township contains the entrance to Carlsbad Cavern, neither the plat nor the accompanying notes mentions the cave's entrance. Surely the land owners named on the plat must have known the entrance well because their holdings are within two miles of it.

The first visitor in the main cave was probably a child. On file in the park office is an 1883 account of a twelve-year-old boy named Rolth Sublett who was lowered by his father from a rope down into the main cave. Although this must have been another hair-raising tale, nothing beyond the lowering is documented. Who then "discovered" Carlsbad's caverns and what consequently became the national park?

DISCOVERY STORIES

Because indigenous man probably knew about the caves 12,000 years ago, it involves a small measure of arrogance to place discovery in the past century. However, the park's principal discoveries, the underground ones, do rightfully belong to this century. So many of the facts and so many of the tall tales about the caves' exploration have been commingled over the years, that picking among it all, one risks placing a yarn in the fact column or calling a fact a piece of fiction. Falling squarely in both columns, however, is a character in the discovery cast who is wrapped equally well in tales and facts: Jim White.

James Larkin White is generally given local but not universal credit for discovering Carlsbad Cavern. Everyone agrees he was among the first to penetrate deeply into the cave and to report his findings. Another point of agreement is that he became a tireless zealot in his support for publicizing the caverns. That he was a guide for two formal expeditions sponsored with outside funds contributed to keeping his name prominent in the discovery documents. Many accounts of White's life at the caverns begin the same way: He was a cowboy who, while riding nearby on his horse, saw an enormous cloud of "smoke" in 1901. His own story says that by the time he reached the origin of the smoke, it was gone. Then he saw the huge opening, but darkness had fallen. White writes that he returned to the cave alone a few days later with a lantern, axe and some rope and lowered himself past the entrance. Several sharp descents later he found himself in the bats' cave and realized he was looking at something better than a gold mine; he was looking at a 100-foot high wall of tons upon tons of bat guano, the premier fertilizer of the time. Turning a different direction and walking far into the cave, he came upon the first of the awesomely beautiful cave features. White reported strange formations in onyx, clear pools of water, long and slender drinking "straws" hanging all around. He was determined that the world should also witness this grand and weird spectacle. He spent much of the rest of his life involved with the cave in one way or another: as a miner, as a guide, and as an employee of the Park Service for a few years.[11]

A differing account attests that his first visit in the cave was with another man, Abajah Long. Long filed a "Placer Location Notice" in the Eddy County Clerk's office in 1903 and was granted the first mining claim, called the "Big Cave Mining Claim." A rift between the two men in later years caused each to maintain a stony silence about the other for the rest of their lives, a rift that may have affected the accuracy of their written accounts of the early days of local exploration and continues to cloud the record.

The issue of discovery credit is further obscured by Frank Nicholson,[12] a writer for the *New York Times*. He wrote in the 1920s about Jim White and his work, but unfortunately Nicholson took personal credit for some discoveries that clearly belonged to White and others and thereby generally cast doubt on everything he would subsequently write about the caves, some of which may have been accurate. Park rangers today continue to debunk Nicholson's boasts and generally consider him to be a fame-hungry charlatan. White himself readily admits that other local residents knew of the cave long before his own entry, and his claim of discovery is based not so much on finding the cave as his explorations of it and the knowledge he gained about it. All these stories of first discovery of this magnificent place, all the claims and counterclaims, boasts and fabrications, were secondary in those days to the real story: bat guano!

By 1903 guano mining was in full swing and cave exploration was a haphazard sideline carried out by White and his friends on Sundays and bad weather days when they did not work the mines. Guano continued to be an economic resource for the area for nearly twenty years. It was much needed in the California citrus industry until the early nineteen thirties when cheaper South American nitrates replaced guano as the fertilizer of choice. During the approximately twenty years of operation, over 100,000 tons of bat guano was removed from the cave, ninety percent of which was there when operations started. [13]

Some people began to hear of the beauty of the caves and wished to see them. To a limited extent during the mining years and to a greater extent after them, Jim White offered tours for $2.00 a day. The only conveyance available then was the guano bucket, which would hold two standing, terrified visitors as it swung wildly back and forth on its 173-foot descent into the pitch blackness. The only light was provided by kerosene lanterns and burns were common. The terror and the rigors of the trip did not deter visitors, however; to see the scenic extravaganza, they began to arrive in bus loads and two by two the world began to see Carlsbad Cavern.

Word spread of the incredible cave. Remarkable early cave photography by a Carlsbad photographer, Ray V. Davis, also publicized the cave to such an extent that the General Land Office (GLO) sent Robert A. Holley, a mineral examiner, to "Carlsbad Cave" in May 1923 to survey it and to report on the feasibility of securing the cave as a national monument.

THE FIRST SURVEY

Holley estimated his work at Carlsbad would take a few days. It took four weeks. He began by running survey lines down the main entrance and establishing a station at a point within 150 feet from the entrance. From there his surveying involved scaling vertical rock faces, descending through narrow openings between jagged rock and dropping up to eight feet with his heavy wooden transit before returning to the "regular" trail, itself very rough. Instead of the predictable cave-tubes he thought he could easily survey, he found an eerily beautiful twisting maze of passages and rooms. He also found himself awestruck and incredulous. His enthusiastic report with an outline map was submitted on May 14, 1923, in prose so flowery that he startled his superiors at GLO. Holley overwhelmingly supported national monument status for the cave, and he also recommended acquisition of additional land for locating the rumored other caves. Further suggestions included building of steps, stairways, walls, and lights; he knew that the onslaught of visitors was coming and that the guano bucket would be of limited value with its two person capacity. On October 25, 1923, President Calvin Coolidge proclaimed the creation of Carlsbad Cave National Monument.

A month earlier, in September of 1923, the United States Geological Survey (USGS) sent one its geologists, Dr. Willis T. Lee, to the cavern to prepare a report for the survey. His manuscript and some of Ray Davis' photographs were published as an article, "A Visit to Carlsbad Cavern," in the January 1924 *National Geographic Magazine.* Public response was such that the National Geographic Society (NGS) began to plan a six month, comprehensive expedition which it would finance with $16,000. Willis T. Lee, who had by this time been named as unpaid Monument Custodian, would lead the expedition.

On Sunday March 22, 1925, Lee entered the main cave accompanied by his daughter Elizabeth, his son Dana, and a USGS surveyor, R. H. Runyan. Jim White served as guide with Carl Livingston of Carlsbad as assistant. Overlapping part of the Lee expedition was the separate one of Vernon Bailey of the United States Biological Survey who was sent to study wildlife in and around the cave.[14]

Frequently using white twine attached from the entrance to the new discoveries, Runyan had a difficult time keeping his sense of direction in the convoluting spaces but finally, after hardship and illness, finished his arduous task of surveying by June 1. His is the first complete map of the cave. He left 48 survey points in red paint upon the walls and large boulders of the main cave, some of which are visible today. Runyan's work, called the Willis T. Lee Map was not only the first but the definitive map of Carlsbad Cavern for over thirty years (fig. 7). In addition to the map, several other benefits came from the Willis T. Lee-NGS Expedition.

An Executive Order of the President set aside many additional public lands for possible inclusion in the park at a later date, a trail was constructed of dirt and guano, and many specimens were sent to the Smithsonian. Lee was politically savvy enough to include important visitors among the privileged few who were allowed to visit the expedition and enter with him. They included the governors of New Mexico and Texas and the President of the National Geographic Society as well as the first National Parks Director, Stephen Mather. These efforts, along with Lee's second article, published in *National Geographic Magazine* in September 1925,[15] contributed substantially to the growing notion that the monument should become a national park.

Carlsbad Caverns National Park was created by an act of Congress and signed by President Herbert Hoover on May 14, 1930, its new plural name reflecting the growing awareness of many more caves in the region. In addition to the original 719 acres set aside with the monument, a provision was made that would allow boundaries to be enlarged as required upon the recommendation of the Secretary of the Interior. Over subsequent years the provision has been exercised numerous times and today the park contains close to 47,000 acres.

PARK DEVELOPMENT

Further exploration within Carlsbad Cavern in the late 1920s located the significant side chambers of Left Hand Tunnel, the Mystery Room, and the New Mexico Room. Ten years later New Cave was discovered in Slaughter Canyon by Park Ranger Tom Tucker. These discoveries notwithstanding, the efforts to prepare the cavern for visitors and to develop it dominate the historical record until the early 1960s. Exploration and mapping of new areas was secondary to the development of known areas. For many years it was mostly park personnel and a few other individuals who conducted

FIRST PUBLISHED MAP OF CARLSBAD CAVERN, 1925

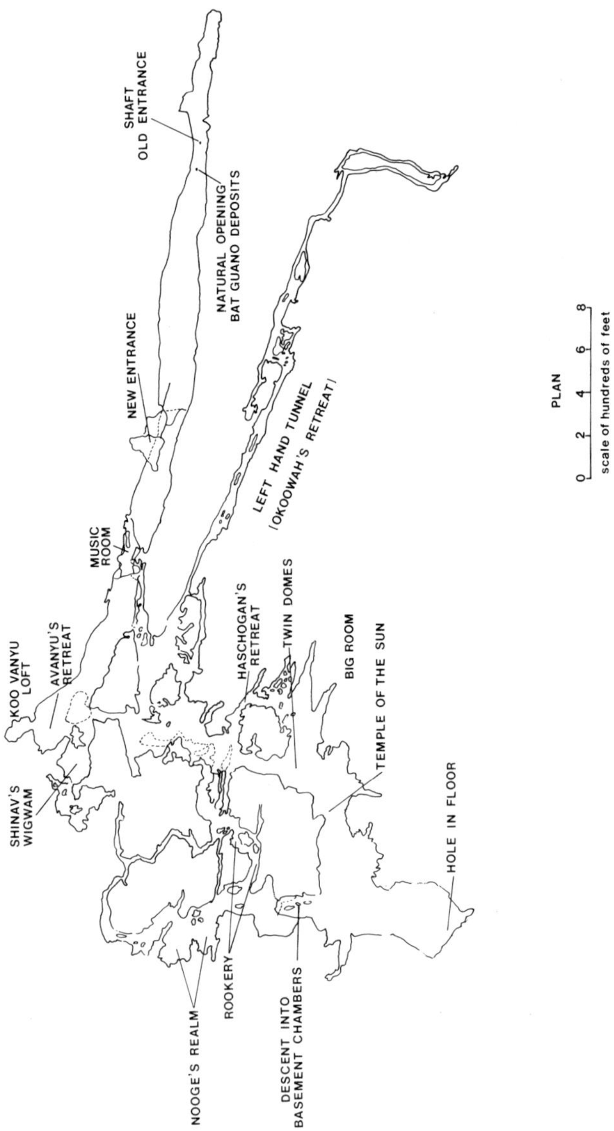

From the first sketch map of Carlsbad Caverns from surveys by Willis T. Lee 1925

Fig. 7.

cave exploration in the park. Understandably, map production was scarce for nearly twenty-five years. Additionally, some park administrators were more supportive of exploration than others.[16] Even without finding new caves, there were enough known but unexplored ones in the region to keep explorers, if not mappers, busy, albeit relatively quiet, for years to come. Ronal Kerbo maintains that finding a new cave represents the dramatic quantum leap of discovery while exploring that cave can represent years of small inglorious steps.[17] For this reason, it may be unfair to say that little exploration occurred at Carlsbad Caverns during the aforementioned years, yet the frequency of exploration stories definitely slowed during a long period of time.

An important aside should be inserted at this point. The history of cave discovery shows that exploration and mapping usually occurred on two different occasions as two separate processes. Scientific explorations and studies came first, and later the surveyors and mappers, frequently from the USGS, arrived to do their work. Philip B. King of the USGS followed Willis T. Lee by a few years and did extensive reconnaissance mapping of the whole caverns area after Lee did the exploration in the usual order of those activities for those times. This order would later change.

The 1920s was surely the pivotal decade in the park's early history of discovery and exploration. In that one decade, some remote caves in southeastern New Mexico, known to very few, become Carlsbad Caverns National Park, known to thousands. After this heady period, the next years would be given mostly to providing for the welfare of those thousands of visitors. Conditions for exploration are vastly different from conditions for visitation and preparations had to be made for the latter. Stairways were built to replace the guano bucket, lights installed, telephone lines brought out from town and concession service provided out of a tent. Later years brought a lunchroom, a paved road from town, a cooling system for the generator, a direct water supply, an elevator, a sewer line, much improved trails and still more tourists. By 1937 one million visitors had been in the park. Concomitant with this interest, USGS produced topographic 1:62,500 sheets of the park in 1939, 1943, 1945, and 1947.

THE NINETEEN SIXTIES AND BEYOND

The next quantum-leap-decade for discovery and exploration is the 1960s. During this landmark decade, several events were significant enough to change forever the exploration of Carlsbad Caverns. One was the discovery in 1966 of several miles of passageway and several rooms including the Guadalupe Room Complex within an area of the main cave thought to have been fully explored. Carlsbad Cavern, the already gigantic cave, became one-third longer! Cave exploration in general began to assume larger proportions as well. Not just at Carlsbad, but throughout the country, a shift was being made from individual exploration to group/agency exploration. This phenomenon of the sixties is perhaps the single most significant event in the history of the exploration and mapping of caves in the United States.

The Guadalupe Cave Survey (GCS) was formed and is considered the first organized group of cavers with a mission dedicated to scientific cave exploration and surveying within the park and the Guadalupe Mountains.[18] GCS did surveys in caves to pinpoint geological formations for the scientists to study later, a reversal of the earlier scientist or explorer-first, surveyor-after process. They worked closely with the National Park Service (NPS) to meet park needs as well. This cooperative effort between the park service and organized groups of cavers continues to the present. GCS eventually joined the Cave Research Foundation (CRF), whose agreement with the National Park Service allowed them to conduct scientific research in any cave in the United States that is on national park land. The CRF produced the first set of detailed quadrangle maps naming formations and rooms for Carlsbad Cavern in 1988.

As a result of technology, another propitious event occurred when exploration was facilitated with more sophisticated survey techniques such as that afforded by the theodolite. In 1964 Tom Rohrer, while a graduate student from Texas A and M University, did the first precision survey of Carlsbad Cavern with the theodolite. Rohrer's surveying of the park's caverns has continued for twenty-five years, while his early theodolite markers remain evident today in the main cave, in some cases not far from Runyan's red paint marks.[19]

THE FUTURE

If only one word could be used to summarize what exploration and mapping of Carlsbad Caverns National Park will entail in the next decade, it would have to be "Lechuguilla." Located on Park Service land, Lechuguilla (named for a prickly plant of the area) is a cave currently measuring nearly 60 miles long but projected to extend even farther. Known locally since 1914, when it was called "Misery Hole," its enormity and uniqueness were actually "discovered" (that misplaced word again) in 1986.

Reaching a depth of 1,501 feet, it is the world's deepest cave and is 47 feet taller than Chicago's Sears Tower. The cave is said to contain other-worldly formations that look like flowers and Christmas trees sprouting from floors and walls, gypsum filaments that are hair-fine and 30 feet long, and many pristine underground pools. Another find is a 245 million-year-old fossil that may be of a heretofore unknown species of snail. At the request of the Park Service, a precision survey in Lechuguilla is designed to provide support for management of wildlife resources. Additionally, it will provide a greater degree of certainty about the location and extent of the major passages of the cave and their relation to surface features in and around Carlsbad Caverns National Park. This will be done by tying to the General Land Office (GLO) and to the geodetic net. [20]

The Lechuguilla Cave Project (LCP) and the CRF, both groups having many scientists who must also be expert cavers, are assuming the physically demanding task of exploring and mapping the virgin cave. To preclude unnecessary extra trips into the fragile biosphere, mapping and exploration are now conducted concurrently; ecological consciousness has entered the process. The former two-tier process of exploration/mapping (or mapping/exploration) has been emphatically discontinued at Lechuguilla. These explorers-mappers are trying not to make the mistakes of the past, mistakes that could irreparably harm this tender place. Among the precautions taken are no carbide lights (found on helmets) allowed in the cave and no waste left behind. Even lint is removed. All survey points must be on breakdown or on non-cave formations. Ecological awareness is a primary consideration of cave exploration/mapping today.

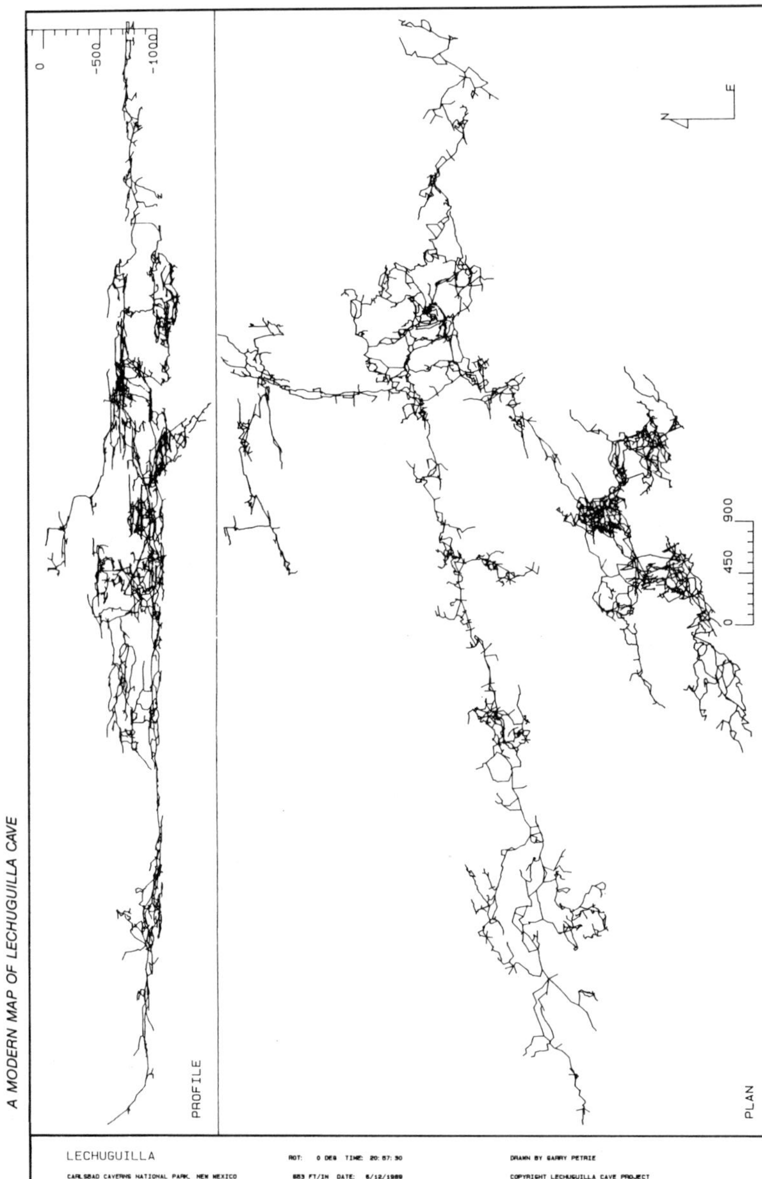

Fig. 8. Courtesy of Lechuguilla Cave Project, Inc. Used with permission.

As new areas are discovered and explored, they are immediately and precisely mapped using computer methods that allow for color-coding depth on the maps, thus creating three-dimensional information on a two-dimensional map. These maps look strange to the average reader, but they are packed with information and are perfectly suited to their task. All features are recorded as they are found, so that a second look is not necessary. Because they are instantly produced, the maps can be finished and published as each Lechuguilla expedition is finished (fig. 8).

When asked "Where can I find new caves?" Ronal Kerbo will reply, "Where no one has looked."[21] Carlsbad Caverns National Park now has seventy-five known caves, but no one assumes there are no others. Perhaps the yet-to-be-discovered caves will prove more grand and mysterious or scientifically significant than the known ones. Known and unknown, they will all require exploring and mapping. Exploration and mapping are never-ending processes in cave environments and this park's odyssey will continue. The only part of the tale that is novel is the techniques of mapping and the new conscientiousness of the cavers themselves. The latest maps may look different, but the dangers and thrills of creating them will never cease, and the story of the exploration and mapping of Carlsbad Caverns National Park will remain a fine ongoing tale.

NOTES

1. Robert Babb, II, James Hardy, and Fritzi Hardy, "The Precision Survey of Lechuguilla Cave," *Professional Surveyor* (May/June 1991): 4-9.
2. Jim Nieland, "Cave Surveying," *The Speleograph* (March 1990): 21-29.
3. B. M. Ellis, "Cave Surveying," in *Science of Speleology* (London: Academic Press, 1976), 1-102.
4. Willis T. Lee, "A visit to Carlsbad Cavern," *National Geographic Magazine* (January 1924): 1-40.

5. Carol A. Hill, *Geology of Carlsbad Cavern and other Caves in the Guadalupe Mountains New Mexico and Texas* (Socorro, NM: New Mexico Bureau of Mines and Mineral Resources, 1987)
6. Ann G. Harris and Esther Tuttle, *Geology of National Parks*, 3rd ed. (Dubuque, IO: Kendall/Hunt Publishing Company, 1983), 554.
7. Bob Crisman, "A Capsule History of Carlsbad Caverns National Park," Unpublished typescript. (Carlsbad Caverns National Park Office, no date), 15.
8. Gale, Bennett T. "Historical Sketch Carlsbad Caverns National Park." Unpublished typescript. (Carlsbad Caverns National Park, Library at Park Headquarters, 1952), 50.
9. Ibid.
10. Ibid.
11. Ruth Caier, *One Man's Dream* (New York: Pageant Press, 1957)
12. Frank Ernest Nicholson, *The Exploration of Carlsbad Cavern* (Wichita Falls, Kansas: Railey Publishing Company, 1930), 48.
13. Willis T. Lee, "New Discoveries in Carlsbad Cavern," *National Geographic Magazine* (September 1925): 301-320.
14. Ibid.
15. Ibid.
16. Bobby L. Crisman, interview with author, Park Headquarters, May 1989.
17. Ronal L. Kerbo, interview with author, Park Headquarters, May 1989.
18. Fritzi Hardy, telephone interview with author, August 1989.
19. Bobby L. Crisman, interview.
20. Robert Babb, II, James Hardy, and Fritzi Hardy, "The Precision Survey of Lechuguilla Cave," *Professional Surveyor* (May/June1991): 4-9.
21. Ronal L. Kerbo, interview.

SELECTED CARTOBIBLIOGRAPHY

Cave Research Foundation. *Carlsbad Cavern. Carlsbad Cavern[s] National Park, New Mexico* [map]. 1" = 200 feet. Albuquerque, NM: The Foundation, 1979.

Cave Research Foundation. *Carlsbad Caverns National Park* [map]. 1:600. 12 sheets. Cedar Falls, IA: The Foundation, 1988.

Cavern Map and Chart Service. *Carlsbad Cavern Trail Map and Guide* [map]. Scale varies. [Carlsbad, NM?]: The Service, 1928.

Lee, Willis T. *Carlsbad Caverns* [map]. Scale varies. Sketch map In: "New Discoveries in Carlsbad Cavern," *National Geographic Magazine*, 3 (Sept. 1925): 302.

Hayes, R. T. *Geology of Carlsbad East quadrangle, New Mexico* [map]. 1:62,500. Geologic Quadrangle Map, GQ-98. Washington, D.C.: USGS, 1957.

Hayes, R. T. and R. L. Koogle. *Geology of the Carlsbad West quadrangle, New Mexico* [map]. 1:62,000. Geologic Quadrangle Map, GQ-112. Washington, D.C.: USGS, 1958.

National Speleological Society. *Carlsbad Cavern. Carlsbad Caverns National Park. Eddy County, New Mexico* [map]. 1" = 50 miles. Huntsville, AL: The Society, 1964.

Owens, Charles H. *Carlsbad Cavern...New Mexico...an underground wonderland* [map]. Scale varies. Sketch map In: *The Los Angeles Times* (Sept. 29,1929).

U.S. Geological Survey. *Carlsbad Caverns National Park, NM* [map]. 1:24,000. Topographic series. Washington, D.C.: USGS, 1934.

U.S. Geological Survey. *Carlsbad, N. Mex.* [map]. 1:62,500. Washington, D.C.: USGS, 1939.

U.S. Geological Survey. *Carlsbad, N. Mex.* [map]. 1:62,500. Washington, D.C.: USGS, 1943.

U.S. Geological Survey. *Carlsbad Caverns West, N. Mex-Tex.* [map]. 1:62,500. Washington, D.C.: USGS, 1940.

U.S. Geological Survey. *Carlsbad Caverns West, N. Mex-Tex.* [map]. 1:62,500. Washington, D.C.: USGS, 1943.

U.S. Geological Survey. *Carlsbad East, N. Mex-Tex.* [map]. 1:62,500. Washington, D.C.: USGS, 1945.

U.S. Geological Survey. *Carlsbad Caverns East, N. Mex.* [map]. 1:62,500. Washington, D.C.: USGS, 1947.

U.S. Geological Survey. *Carlsbad, N. Mex.* [map]. 1:62,500. Washington, D.C.: USGS, 1971.

U.S. Geological Survey. *Carlsbad East, New Mexico* [map]. 1:24,000. Washington, D.C.: USGS, 1985.

U.S. Geological Survey. *Carlsbad West, New Mexico* [map]. 1:24,000. Washington, D.C.: USGS, 1985.

REVEALING THE SAPPHIRE SECRET OF THE CASCADE RANGE: EXPLORATION AND MAPPING OF CRATER LAKE NATIONAL PARK

Peter L. Stark

University Library
University of Oregon

The landscape surrounding Crater Lake gives no indication that a grand volcanic monument of the Cascade Range lies just over the horizon. Approaching from the west, thick stands of Douglas fir and western hemlock are left behind as one moves upward along a steep and curving road to higher elevations where lodgepole and white pine cover the slopes. The forest at the crest of the Cascades is widely spaced and dominated by groves of stately mountain hemlock. From the east, the way to the crest of the Cascades follows a remarkably straight and easy grade where ponderosa and sugar pine forests flourish at the lower elevations before giving way to the mountain hemlocks. Between the pointed, bare summit of Mount Thielson and the north rim of Crater Lake lies a broad parklike tableland without streams of any size and studded with volcanic buttes. It is difficult here to determine the location of the crest of the Cascade Range. Ridges and valleys mark the passage to the southern rim of the lake, where watercourses are no larger than creeks. There is nothing in the approach to Crater Lake that suggests that this portion of the Cascade crest is any different from others in southern Oregon, no clue of the vast lake that soon will come into view

The startling sapphire blueness of Crater Lake and its majestic silence have greeted and inspired countless visitors to its rim. Early prospectors and travellers bestowed several names on Crater Lake unaware of the fact that it had previously been visited and named by others. Names such as Deep Blue Lake, Lake Majesty, or Mysterious Lake all indicate common impressions and shared experience; for native Americans, the lake held a special meaning.

PRE-HISTORY

The many native American legends about Crater Lake leave no doubt that its existence was well known to the Klamath, Modoc, and Rogue River tribes that inhabited the region. The spirit and character of these legends confirm that the wonder and mystery of Crater Lake expressed by white visitors were also felt by the native inhabitants. Most early references to the native Americans' relationship to Crater Lake accented their "fear" of the lake and neglected to examine their feelings of awe and reverence for the high and sacred place. William Gladstone Steel, the man chiefly responsible for having the area set aside as a national park, wrote in 1890 that "the Indians of Southern Oregon have known of it [Crater Lake] for ages, but until recently none have seen it, for the reason that a tradition, handed down from generation to generation, described it as the home of myriads of sea devils, or, as they were called, Llaos; and it was considered certain death for any brave to look upon it."[1]

In 1885 Chief Allen Davey of the Klamaths told Steel a legend of Crater Lake that explained why the Klamaths avoided the lake, why they willingly served as guides but would halt before the rim was achieved and proceed no farther. Long before white men came to the West, a band of Klamath Indians were hunting in the mountains when they suddenly found themselves on the rim of a deep crater. Startled, they gazed in silence at the lake with water bluer than the sky. Believing in their hearts that this was the home of the Great Spirit, and knowing that punishment came to those who look upon places he held sacred, they moved quickly away from the rim to camp far away.

One hunter, however, felt a strong urge to return. The next evening he climbed back to the rim and camped there for several nights. He finally found enough courage to climb down to the water's edge, to swim and camp, and although he heard strange voices arise from the lake he was unafraid. Soon he became stronger than any other member of the Klamaths. He confessed his secret and soon others began to seek the lake's powers. They, too, became healthier and stronger.

One day, the hunter that had first dared to visit the lake killed one of the creatures that lived in its blue waters. He was immediately set upon by untold numbers of angry Llaos who carried the hunter to a high cliff above the lake. There, they cut his body into small pieces and threw them into the waters below to be devoured by other Llaos. This would be the fate of anyone who ventured too close to the waters of Crater Lake.[2]

For the Indians who lived on the eastern slopes of the Cascades, Crater Lake, or Gay-was, had its origins in the battles between good and evil. The creation legend closely resembles events that geologists believe actually occurred. In 1865 a young soldier stationed at Fort Klamath, William M. Colvig, listened to Chief Lalek of the Klamaths recount the origin of Crater Lake. Lalek told Colvig that long before Crater Lake was formed there existed a high mountain in its place. There was an opening in the high mountain through which the Chief of the Below World passed from his home to the outside world. One time, when the Chief of the Below World was on earth, he saw the daughter of the tribal chief and soon fell in love with her. He promised the maiden eternal life if she would marry him but she refused. He pressed his suit with the tribal elders but was once again refused, and compounding the offense, the elders sought to hide the girl.

Enraged, the Chief of the Below World vowed to take revenge on the tribe by destroying them with the curse of fire. He emerged from the top of his mountain in a wild fury. The Chief of the Above World saw what was about to happen and quickly descended from his home in the sky to settle on top of Mount Shasta. The two spirit chiefs battled. The earth shook and mountains crumbled. Hot rocks and burning ash fell thickly from the sky. Day turned to night. Fire consumed the forests. The people fled in terror before the fire before finding refuge in the waters of Klamath Lake. There, they prayed to the Chief of the Above World to save them from destruction.

The tribal elders knew what had caused the curse of fire. They sadly concluded that only a living sacrifice would appease the wrath of the Chief of the Below World and that they themselves must be the ones to offer their lives. The voice of the Chief of the Above World boomed from the top of Mount Shasta saying that the elders had reasoned well, and he urged them to act quickly to prevent the total destruction of the people. Two of the oldest and most revered elders left the waters of Klamath Lake and, with burning torches, began the ascending the mountain. Reaching the top of a cliff which extended over the entrance to the Below World, they raised their torches over their heads and leaped into the burning crater.

The Chief of the Above World saw the brave deed. Again the mountain shook and the Chief of the Below World was driven back to his home underground, the mountain falling in upon him. When the sun rose the next day the mountain was gone, no longer challenging the nearby peak sacred to the Chief of the Above World, Mount Shasta. For many years after, rain fell and filled the great opening that was created when the mountain collapsed. Peace once more filled the earth. Never again did the Chief of the Below World come from his home to frighten the people.[3]

It was once believed that native Americans had not arrived in what is now the state of Oregon early enough to have witnessed the great volcanic cataclysm which gave birth to Crater Lake. But radiocarbon dating of artifacts found at Fort Rock Cave, fifty-five miles northeast of Crater Lake, indicates an Indian presence in the area well over 9,000 years ago. The native American legends describing the fiery past of Crater lake were undoubtedly inspired by eyewitness accounts.[4]

Crater Lake lies inside the collapsed top of an ancient volcano called Mount Mazama. The bowl-shaped depression holding the lake is not a mere summit depression, the definition of a crater, but a caldera. The ancient Mount Mazama, one of a series of volcanic cones extending about six hundred miles from Lassen Peak in California to Mount Garibaldi in British Columbia, began to form more than 400,000 years ago at the crest of the Cascade Range. Geologists believe that Mount Mazama was never a symmetrical volcano but was composed of several overlapping cones, each active for a comparatively short period. One of the oldest and most prominent of these cones was Mount Scott, now standing high and alone only two miles from the caldera rim. Mount Mazama probably rose to a height of between 11,000 and 12,000 feet and, similar to other high volcanic peaks in the Cascade Range, experienced several glacial advances and retreats on its flanks.

About 6,600 years ago, a series of culminating eruptions began. Immense lava flows and explosive eruptions drained the great underground reservoir of magma. New vents opened along the sides of the volcano allowing huge avalanches of glowing pumice, ash, and boulders to race twenty to thirty miles from their source, which further drained the magma reservoir underneath the volcano. Finally, the roof of the magma chamber caved in along a semi-circular zone of vents. The entire culminating eruption and the formation of the caldera occurred simultaneously, probably during a period of only a few days. The high peak of Mount Mazama, a familiar landmark to native Americans, had vanished. Left behind was a smoking caldera, five to six miles in diameter and four thousand feet deep. The forests for miles around were laid waste, and the caldera together with the surrounding valleys and slopes continued to emit acrid fumes for many years as heaps of hot ash and pumice slowly cooled. The collapse of the summit was eccentric with respect to its base for the top of Mount Mazama lay well south of the center of the caldera. The southern rim of the caldera, with its glacially carved U-shaped valleys left headless after the collapse of the volcano, rises much higher than its northern counterpart.

Activity did not completely cease as volcanic domes erupted from the floor of the caldera. The oldest rises from the center of the lake to within four hundred feet of the lake's surface. Wizard Island to the west of this oldest dome is built on a much larger base and represents, like an iceberg, only the uppermost seven hundred and fifty feet on a mass of lava over 2,500 feet high.

Crater Lake, formed almost entirely by infalling rain and snow, is the deepest body of freshwater in the United States, having been sounded to a depth of 1,932 feet. The level of the lake varies only slightly with the seasons and the amount of water added each year is about the same as the amount lost by evaporation and seepage. No streams flow from the lake; it is entirely enclosed by a rim nine hundred to two thousand feet above the water. Nor has any lake water been found to issue from any of the springs on the flanks of the now decapitated Mount Mazama.

Both the Klamath legend and the geologic interpretation of Crater Lake describe the transformation of the monstrous into the sublime. Even for the veteran visitor, Crater Lake never fails to impress and inspire.

EARLY EXPLORATIONS IN SOUTHERN OREGON

Many explorers, trappers, and surveyors passed through southern Oregon in the first half of the nineteenth century but none encountered Crater Lake. The reasons for this have much to do with the isolation of Crater Lake and with the purposes that the individuals and parties carried with them into the wilderness.

The first Euro-Americans to enter southern Oregon were probably trappers working for the Hudson's Bay Company. Coming from Fort Vancouver in the north, they travelled down the west side of the Cascade Range as far south as the Rogue River drainage. Later more organized trapping expeditions were launched from Fort Vancouver, one in 1820 led by Thomas McKay to the Willamette Valley and another in 1826 under Alexander Roderick McLeod along the Oregon coast. In the same year, 1826, John McLoughlin, chief factor of the Hudson's Bay Company, detailed a large expedition to explore further and assess the economic potential of southern Oregon west of the Cascades. The party was joined by David Douglas, a British botanist, who collected and identified many species of plant and animal life in the Pacific Northwest, notably the Douglas fir and Douglas squirrel.

Meanwhile, Chief Factor McLoughlin had not overlooked the eastern slope of the Cascades. Late in 1826 McLoughlin sent a trapping and exploring expedition into the Snake River Country, one of several expeditions headed by Peter Skene Ogden. The 1826-1827 Snake Country Expedition visited the Klamath Lakes area in December 1826, but a shortage of supplies and the poor weather directed the party to the south and then west. Ogden sighted Mount Shasta before entering the heart of the Rogue River drainage on the western slope.

The American Jedediah Smith and a party of eighteen men entered southern Oregon from California in 1828. Smith followed the coast trapping beaver with great success as he marched north. The arduous journey toward his destination, Fort Vancouver, met with tragedy near the mouth of the Umpqua River, when in July of 1828, his party was ambushed by Indians. Fourteen men were killed. Four survivors, including Smith, made their separate ways to Fort Vancouver and were assisted in the recovery of their pelts and other belongings by John McLoughlin.

The participants in the Hudson's Bay Company sponsored excursions into southern Oregon as well as Jedidiah Smith's party were probably not tempted to investigate the high country as beaver, their chief objective, were plentiful along the untrapped rivers and streams of the lowlands. Also, confrontations with the Indians dictated short excursions into the area. Of all these early entrants into southern Oregon, Ogden drew the nearest to Crater Lake, but approached the region in December...certainly not the best month of the year to explore the stormy high Cascades.

Two famous American explorers, Lieutenant Charles Wilkes and John C. Fremont, dominate the exploration of the area in the 1840s. While in Fort Vancouver in 1841, Wilkes gathered a party of thirty-nine soldiers, sailors, botanists and other scientific men, guides and hunters and ordered this detachment to explore the land route between the Columbia River and the Sacramento in California. Specifically, they were to look for any large streams in the interior that could possibly indicate harbors at their mouths. Wilkes himself planned to sail to San Francisco where he would await the detachment's return. Under the command of George F. Emmons, the exploring party encountered constant Indian harassment. Thick smoke from forest and brush fires deliberately set by Indians plagued their progress south. Their difficult experience in southern Oregon resembled a exodus rather than an organized exploring expedition as the constant confrontations with the Indians and sickness among the men made the party desire California and the return to tidewater all the more. The passage south from the Columbia River to Sutter's Fort in California, over 550 miles of hardship, took forty-three days (September 7 to October 19, 1841) and afforded little time for much real exploration. However, the route they blazed south along the west side of the Cascades, the most direct to California, would be followed by many travellers after them.

During the 1840s John C. Fremont entered Oregon twice, once on his 1843-1844 expedition and again in 1846; on both occasions, he was exploring the eastern slopes of the Cascade Range. Fremont's 1843-1844 expedition took him west and into Oregon by generally following the Oregon Trail. Stopping at The Dalles, he sent a party to Fort Vancouver, thus technically connecting his expedition with that of Wilkes two years earlier. Regrouping, he set out from The Dalles heading south along the Deschutes River, reaching Klamath Marsh, which he mistakenly thought was Klamath Lake, his intended destination, on December 10, 1843. From his camp on the marsh, Fremont observed that "immediately west, were several snowy knobs, belonging to what we have considered a branch of the Cascade Range."[5] The most prominent "snowy knobs" sighted were Mount McLoughlin (known at the time as Mount Pitt) and Mount Scott, the latter being the closest to his encampment, only twenty-five miles distant and located on the rim of Crater Lake.

Even though the Klamath Indians knew of Crater Lake and the surrounding country, they were quite uncooperative when it came to communicating knowledge of their homelands when requested to do so by Fremont. "From Klamath Lake [Marsh], the further continuation of our voyage assumed a character of discovery and exploration, which from the indians here, we could obtain no information to direct . . ."[6] Fremont turned east from Klamath Marsh visiting and naming Summer Lake and Lake Abert before resuming his winter journey south into Nevada and California. Like Ogden's 1826-1827 Snake Country Expedition, the Fremont expedition of 1843-1844 arrived in the vicinity of Crater Lake in December, too late to explore the high Cascades.

On his 1845-1846 expedition, Fremont again visited the Klamath Lakes basin hoping to link up with his Klamath Marsh trek of 1843. In May of 1846, his party camped on the west side of Upper Klamath Lake, but a messenger overtook Fremont with news of an impending war with Mexico. The exploring party turned south and retraced its path back to California, precluding further exploration of southern Oregon. Unlike Fremont's first expedition into the Klamath Lakes region, this time the season was much more conducive to a visit to the high Cascades, but looming events in California persuaded Fremont to leave southern Oregon.

Further and more detailed explorations were made by Robert Williamson and Henry Abbot, both Army engineers who had been ordered to survey two possible railroad routes from San Francisco Bay to the Columbia River, one on the east, the other on the west side of the Cascade Range, as part of the Pacific Railroad Surveys. In August of 1855, Williamson and Abbot moved into southern Oregon from California on the east side of the Cascade Range surveying a route along the eastern shore of Upper Klamath Lake. The engineers also investigated potential railroad passes across the Cascades near the Middle Fork of the Willamette River, the McKenzie River, and the Clackamas River. They took the same route along the west side of the Cascades as taken by many trappers and immigrants south to California. A wagon road crossing the mountains from the Rogue River drainage to Klamath Lake in southern Oregon had already been blazed by Jesse and Lindsay Applegate in 1846. Called the Southern Emigrant Route, it was not investigated by the Williamson and Abbot survey but it was located on their map of the region.

Lieutenants Williamson's and Abbot's two-sheet map of "Routes in Oregon and California" (Map No. 1 *From San Francisco Bay to the Northern Boundary of California* and Map No. 2 *From the Northern Boundary of California to the Columbia River*), dated 1855 and published in volume 11 of the quarto edition of the *Pacific Railroad Report* in 1859, were the best maps of the area to that time. Certainly they were the most honest, leaving great land areas blank and superimposed with the word "unexplored." A good deal of the Cascade Range in southern Oregon, including the area occupied by Crater Lake, was left blank with tentative rivers leading out of the blank spaces. Over this entire area, Williamson and Abbot noted, "Heavily timbered ridges separated by immense ravines."

The Williamson and Abbot maps of Oregon and California, along with all other cartographic material to come from the Pacific Railroad Surveys, were combined with other official data from the General Land Office and the office of the Army's Topographical Engineers by Lieutenant G. K. Warren in his general map of the West, included in the *Pacific Railroad Reports.* In his compilation of this general map, Warren represented only such areas that had been examined and explored. "I have," said Warren of his 1857 general map of the West, "according to my instructions, represented on it only such mountains as are known. In some instances where exploration of known ranges was imperfect, they have been but faintly engraved, the word 'unexplored' placed near them."[7] He had also taken care to resolve discrepancies in geographical positions. Carl I. Wheat has said that the resulting map "was unquestionably the most accurate that had been made, up to this date" and further that "it is believed that the information on which this map was based was sufficient to afford a reasonable over-all picture of the American West, that its compiler made effective use of that information, and that subsequent efforts in the way of maps may properly be deemed merely the filling in of detail."[8]

Needless to say, Warren's map of 1857 did not show the location of Crater Lake, nor had any other map by this date. His map seems an appropriate landmark with which to conclude this period of exploration and mapping of the southern Oregon region. However, the lake had actually been discovered four years earlier by a group of frontiersmen who helped to "fill in the detail," as Wheat put it, on maps of the West...the prospectors.

DISCOVERY AND EARLY VISITORS

Gold had been discovered in southern Oregon in 1851. By 1852 rich diggings were reported in and around Jacksonville which drew many prospectors from California. One of these, John Wesley Hillman, an original "49er," gradually worked his way north to Oregon from the spent placers of the Sierra.

In the spring of 1853, a group of eleven Californians secretly, they thought, entered Jacksonville and tried to look like anything but a prospecting party. They quietly and quickly bought supplies and mules and headed out of town the next day. Hillman, however, had learned from one of the members of this party, after a good deal of drinking, that the prospectors had come to Oregon to seek the legendary "Lost Cabin Mine." New information had been found, said Hillman's informant, that would enable them to find the fabulously rich mine.

It did not take Hillman long to gather another prospecting party of his own and set out after the Californians, intending, if their search was successful, to stake out adjoining claims. The Californians became aware of the party to their rear and therein ensued several days of back-trailing, camping in impossible places and creek-bed walking to throw Hillman's party off their track. The only result of these diversions was that provisions ran low and the parties became disoriented. Returning from a hunt, Hillman encountered a lone Californian, and after informing him that he knew what they were after, Hillman proposed that the two parties join forces. This arrangement was agreed to by the Californians. Half the men were sent back to Jacksonville due to the shortage of provisions; the other was to go on searching for the mine. Now deep in the Cascades and all intent on one object, the landmarks to the Lost Cabin Mine still could not be found.

On June 12, 1853, Hillman was riding a mule up a gently sloping mountain when the animal stopped. He became aware that the slope had vanished and that there was no summit to this mountain. He was soon staring down at an unbelievably large and blue sheet of water. The rest of the prospecting party was gathered, and by a vote, Deep Blue Lake was chosen for a name. Hillman later claimed that "the finding of Crater Lake was an accident, as we were not looking for lakes; but the fact of my being first upon its banks was due to the fact that I was riding the best saddle mule in southern Oregon."[9] The prospectors returned to Jacksonville without finding the Lost Cabin Mine but told others about their discovery of a large and beautiful lake. "Stranger to me than our discovery," wrote Hillman many years later, "was the fact that after our return I could get no acknowledgement from any Indian, buck or squaw, old or young, that any such lake existed; each and every one denied any knowledge of it, or ignored the subject completely."[10] Hillman emerged from his futile search for the Lost Cabin Mine quite disoriented and perhaps sought directions from the Indians back to his Deep Blue Lake but, as the Indians considered the place sacred, they were naturally reluctant to discuss its location with Hillman. Because southern Oregon lacked a newspaper to communicate his discovery and the region's growing concern with Indian hostilities, the news of Hillman's find was largely ignored. The absence of any earlier documentation has led to the conclusion that Hillman was the first white man to look upon Crater Lake. Hillman was fortunate to be alive, at age seventy, to see Deep Blue Lake designated Crater Lake National Park.

Nothing further was heard of Crater Lake for nearly ten years after Hillman's unsung discovery. Then on October 21, 1862, six miners discovered the lake unaware of the 1853 encounter. The men were returning to winter in the Rogue River Valley after leaving central Oregon's Granite Creek mines in the John Day Country. Crossing the Cascades and searching for a camping site on high ground from which to view the surrounding terrain, a common method of navigation in early days, the party suddenly and unexpectedly stumbled upon the lake. One of the party, John W. Sessions, a few weeks later wrote, "Before us, and at our feet, lay a large lake, encircled on all sides by steep and almost perpendicular bluff banks, fully as high as that we were standing upon . . . The waters were of a deep blue color, causing us to name it Blue Lake."[11] Sessions' account of his discovery was the first to appear in print. It was also significant in that the article in Jacksonville's *Oregon Sentinel* also provided an approximate location of the lake. "It lays about one mile west of Mount Scott; fifteen miles south of Diamond Peak, and eighty miles northeast from Jacksonville."[12] Actually, it lies about sixty miles south of Diamond Peak, but the other directions and milages were quite close.

The establishment of Fort Klamath north of Upper Klamath Lake on the east side of the range in 1863 was directly responsible for increased exploration of southern Oregon Cascades in the later half of the 1860s. Fort Klamath's small garrison was charged with protecting emigrant wagon trains from Indian harassment and, more importantly to the history of Crater Lake, the soldiers were instructed to build new roads and improve old trails in the region. The need to connect Fort Klamath with Jacksonville, the largest settlement on the west side of the Cascades, with an adequate wagon road led to the third discovery of Crater Lake.

Under the command of Franklin B. Sprague, soldiers from Fort Klamath moved northwest following Annie Creek south of the lake, cutting timber in the path of the new wagon road in the summer of 1865. Two hunters attached to the road building party excitedly rode back to camp on August 1 and reported to Captain Sprague that they had chanced upon a vast blue lake. Sprague decided to visit the place as soon as possible, but he did not have the opportunity to fulfill his resolve until August 24. In his search for the lake, Sprague and a small detachment of soldiers were joined by several civilians from Jacksonville who had arrived in the area to view the progress of the wagon road and to search for the wondrously blue lake that had existed for them in rumor only. At nine o'clock in the morning of August 24, 1865, the party of soldiers and civilians stood on the rim of Crater Lake. Sergeant Orson A. Stearns, captivated by the beauty of the water, decided to pick his way down the slope to the lake's surface, and in a few hours, became the first in recorded history to reach the shoreline. Others of the party followed, who, when reaching the water, suggested that Stearns have the honor of christening the lake. However, Captain Sprague put forward the name Lake Majesty to the satisfaction of all at hand. Sprague returned to Fort Klamath and wrote of this third discovery of Crater Lake, which appeared in the *Oregon Sentinel* on September 9, 1865.

On September 3 in the same year, a group of eleven hunters on their way to Diamond Peak arrived on the west side of Crater Lake and were able to trek down to the shoreline. A large party of civilians, including two women, from Fort Klamath visited the lake in the fall of 1865. Upon reaching the lake Annie Gaines (Annie Creek) and Mrs. O. T. Brown became the first women to make the journey down to the water's edge.

The wagon road between Fort Klamath and Jacksonville was completed in 1866. The road passed within three miles of Crater Lake's rim, making a short excursion to viewpoints attractive to all who travelled upon the road. But the journey up from the valleys, especially the western approach to the crest of the Cascades, could in no way be considered an easy passage or a trip to be spontaneously undertaken. The editor of Jacksonville's *Oregon Sentinel* newspaper, James A. Sutton, organized a well publicized expedition from the town to Crater Lake in the summer of 1869. Preparations were made to transport at least a score of men, women and children, their supplies, and a knock-down wooden boat up the western approach to Crater Lake. After a difficult ascent up the winding narrow road, the party reached Crater Lake early in August. Several firsts resulted from Sutton's expedition of July/August 1869, namely: the first boat navigated on Crater Lake; the first sounding made of the lake's bottom; the first to reach the summit and then to name Wizard Island; and, in the written reports of Sutton's expedition that appeared in the

Oregon Sentinel for August 28, 1869, the first to use the name Crater Lake.[13]

With an access road in place and with several printed reports in circulation all describing the beauty of the lake, sojourns to Crater Lake among local inhabitants became more frequent and thus less noteworthy. The lake become known to outsiders only gradually. The Modoc War of 1873 brought travellers, writers and later, for the trial of the Modoc leader, Captain Jack, officials from Portland, Oregon, into the region. S. A. Clarke wrote of his 1873 trip to the lake for the December issue of the *Overland Monthly* that same year and Francis Fuller Victor, one of H. H. Bancroft's ghost historians, described her 1873 travels to Crater Lake in her book *Atlantis Arisen.*[14] Peter Britt, southern Oregon's most famous artist and photographer, took the first photographs of Crater Lake in 1874. Despite these contacts, Crater Lake remained a locally popular but, to the outside world, a largely unknown natural wonder for nearly three decades after its discovery by Hillman in 1853.

The maps of the period from 1857 to 1873 reflect the limited awareness of Crater Lake and a limited understanding of the geography of the southern Oregon Cascades. To explain this state of affairs, it must be remembered that the four post-Civil War surveys of the West, which were bringing great amounts of information about the geography, geology and other physical characteristics of the frontier before the American people and the Congress, did not investigate, photograph, or map the Pacific Northwest. Scientific exploration and mapping by the federal government would have to wait until the 1880s and the United States Geological Survey.

The commercial mapping produced in the east by the firms of Mitchell, Cram, Colton, Johnson, and others continued to use the 1857 Warren map of the West as their principal base throughout this period. Crater Lake would not be represented on their maps until better information, usually derived from state maps produced by the General Land Office, became available. Even the landmark western maps of the 1860s produced by William H. Knight for Hubert Howe Bancroft failed to show Crater Lake, probably due to the absence of detailed and authoritative maps from governmental sources. The General Land Office maps of Oregon published during the 1860s and into the 1870s were not as concerned with areas of the high Cascades as they were with showing the extension of the public land survey into arable areas and thus consistently rendered high mountain areas with impressionistic hachures without any real accuracy. This would all change in the mid-1870s.

All too frequently in modern times, wars have fostered advances in the accuracy of maps. So it was with the Modoc War of 1873. The war focused unprecedented attention of westerners on the Modoc Country of southern Oregon and northern California. This attention, without a doubt, led to the publication, in Portland, of the first map to show Crater Lake. Entitled *Habersham's Sectional and County Map of Oregon compiled from the most recent U.S. Government Maps & Surveys*, this map was published by "J. K. Gill & Co., Booksellers and Stationers, Portland, Ogn." in 1874 and was the work of Robert A. Habersham, a civil engineer and draftsman responsible for several other noteworthy maps of Oregon and the Pacific Northwest (fig. 9). The map shows contiguous areas in California necessary to portray the significant events of the Modoc War, such as the spot where Captain Jack was finally captured. A note in the bottom left margin reveals the reason for the leap in accuracy that this map represents in the portrayal of southern Oregon: "The Author [Habersham] acknowledges the receipt of valuable information from Col. John Mendenhall. 4th Artillery, U.S.A. and Jesse Applegate Esq. respecting the 'Modoc Country' and the vicinity of Klamath, Wright and Goose Lakes." Colonel Mendenhall probably supplied Habersham with information on the "Modoc Country" and the events and movements of the war, while Jesse Applegate communicated his wide knowledge of the region in general to the Portland cartographer.

Jesse Applegate had become one of the most venerable citizens of southern Oregon by the time of the 1873 Modoc War. He and his brother Lindsey had blazed the Southern Emigrant Route in 1846 across the southern Oregon Cascades which later became known as the "Applegate Trail." Jesse had entertained Francis Fuller Victor in 1873 with Oregon Indian lore and had guided her to the rim of Crater Lake. Applegate was an exceptionally fine source of information on the geography of southern Oregon.

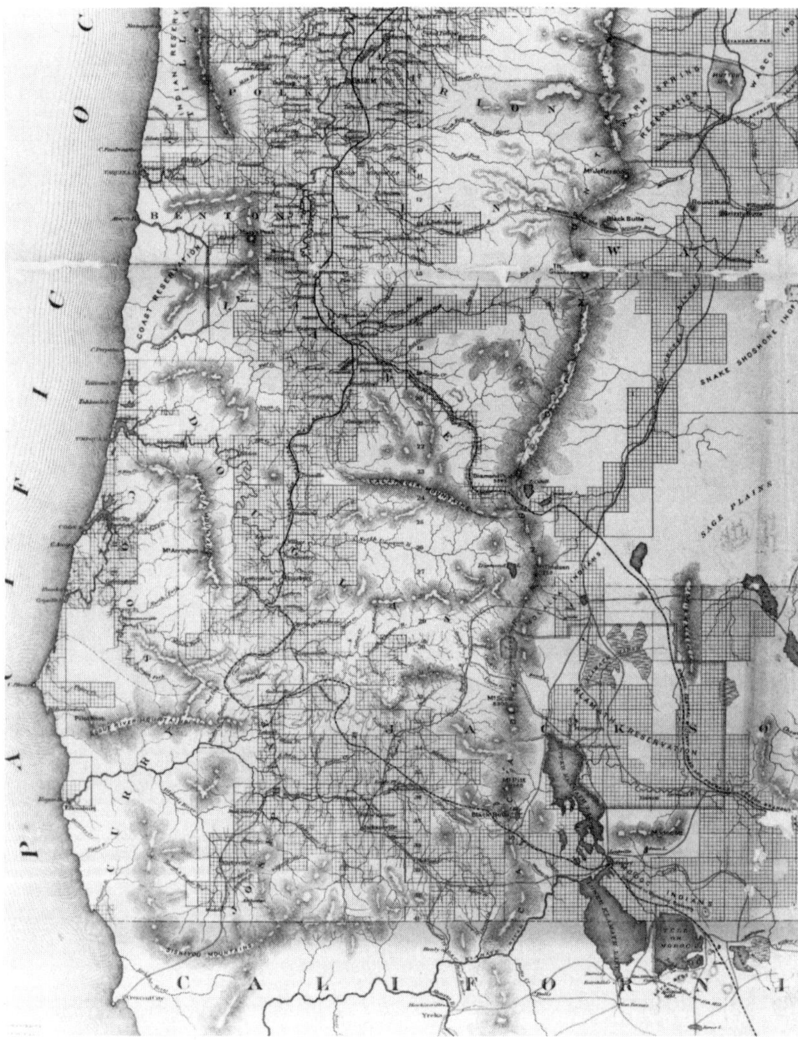

Fig. 9. Southwest portion of *Habersham's Sectional and County Map of Oregon* . . . by Robert A. Habersham, 1874.

Crater Lake is named on the Habersham map and appears at the summit of the Cascades, shifted a bit to the northeast but remarkably near to its actual geographical position. Habersham's 1874 map, drawn to a scale of one inch to twelve miles, shows an elongated Crater Lake about three miles across east to west and six miles north to south, quite close to its elliptical five to six mile diameter. The Jacksonville-Fort Klamath road is shown as running south of Crater Lake and, significantly, no creek or river is shown to have the lake as its source. However, one glaring and inexplicable error did exist, one that would be copied onto many other maps for the rest of the 1870s and well into the 1880s. Mount Scott is depicted as being some twelve to fifteen miles to the south of Crater Lake, not in its correct position on the east rim of the lake. Altogether, Habersham's 1874 map was remarkable for its accuracy and that it was published locally in Portland. Its publication was due to the increased public attention focused on the region's Modoc War. That Crater Lake appeared on this map at all is owing to the fact that Habersham relied on indigenous sources.

The General Land Office's 1876 map of Oregon was the agency's first to show and name Crater Lake, but it did not represent an advance over Habersham's map of 1874, excepting of course the current information provided concerning the extension of the public land survey. Crater Lake is much smaller than on Habersham's map and Mount Scott is shown to be far south of the lake. This map enjoyed a much wider distribution than the earlier Habersham map; its depiction of a smaller lake with Mount Scott placed a bit further to the south of Crater Lake than Habersham's was the example that was followed by other mapmakers. Eastern commercial cartographers, such as Augustus Mitchell, included Crater Lake on their separate maps of Oregon and in atlases for the first time beginning in 1880, showing an undersized lake with Mount Scott out of place. Placing these two features in their correct geographical positions was accomplished by the local office of the Army Engineers in 1881.

One of only five maps with enough significance to be included by Carl I. Wheat in his cartobibliography of the Transmississippi West for the decade of the 1880s was the 1881 *Map of the Department of the Columbia* which was "projected and compiled at the Engineer Office,

Department of the Columbia" at Vancouver Barracks in Washington Territory. This map incorporated new information that had reached the Army's Engineer Office during the 1870s and was the first to depict Crater Lake and Mount Scott side by side. The General Land Office quickly employed this landmark map in the compilation of its state maps of Oregon from the early 1880s onward. Israel Cook Russell's report of his 1882 geological reconnaissance in southeastern Oregon, published in the fourth annual report of the Geological Survey, plainly uses this map, drawn by Lieutenant Thomas W. Symons and Alfred Downing, as its base.[15] Commercial publishers, including Rand McNally and remarkably J. K. Gill of Portland, who had published Habersham's breakthrough map in 1874, were slow to adopt the Army Engineer Office's 1881 map in their cartographic products.

By 1881 Crater Lake was accurately represented on reliable maps issued by official sources. Commercial cartographers were slow in including new information about Crater Lake until the middle of the next decade. Soon, however, the Geological Survey was to explore Crater Lake, survey its region with greater accuracy, and investigate its geology.

SCIENTISTS AND CIVILIANS

Two significant events occurred in the 1880s which would bring Crater Lake out of its isolation and onto the national scene. The first involved the decision early in 1882 of John Wesley Powell, the director of the United States Geological Survey, to first map and then to conduct a geological reconnaissance of the Cascade Range in northern California and Oregon. Powell assigned one of his most talented geologists to the task, Clarence E. Dutton, who had just completed work on his monograph entitled *The Tertiary History of the Grand Canyon District*, perhaps the most renowned early publication of the Geological Survey.[16] The second event, no less important than the first, was the visit in 1885 of William Gladstone Steel to Crater Lake. Upon viewing the lake, Steel embarked on a seventeen year crusade to have Crater Lake established as a national park, a task assisted through the support of federal scientists.

The Cascade Range extending from California's Lassen Peak to British Columbia had not been investigated or mapped by any of the four great post-Civil War surveys of the West, although George M. Wheeler's survey did explore the eastern approaches of the range in California. Into the unsurveyed Cascades, Director Powell sent Gilbert Thompson to begin mapping the volcanic mountain range north of the fortieth parallel in 1882. Mapping would proceed along Powell's original plan for a uniform topographic map of the United States: a scale of 1:250,000 or four miles to the inch, vertical relief shown by contours drawn at intervals of two hundred feet in mountainous areas, and publication in atlas sheets of one degree of longitude by one degree of latitude. Clarence E. Dutton was put in charge of geologic studies, but convinced Powell that he should first see active volcanoes before he investigated the volcanic Cascades. Further, he explained, geologic work could not begin until the topographic mapping had made some progress. Instead of conducting field work in the Cascades in 1882, Dutton spent several months in the Hawaiian Islands in preparation for his domestic assignment.

Mapping continued under Gilbert Thompson during the 1883 field season and in the same year, the geologist Joseph Silas Diller joined the Geological Survey as assistant to Dutton. Dutton remained in Washington, D.C., to complete his memoir of the volcanoes of the Hawaiian Islands[17] and wrote in his annual report for 1883-1884, "I directed my assistant, Mr. J. S. Diller, to make a preliminary reconnaissance of that portion of it [the Cascade Range] which lies between the head of the Sacramento Valley and the Columbia River . . . After visiting Mount Shasta he proceeded to the eastern base of the range, gradually working his way to Fort Klamath. From that post he visited the more striking volcanic piles in its vicinity, and also the very interesting locality known in the vicinity as Crater Lake."[18] Diller and Ensign Everett E. Hayden, United States Navy, conducted a preliminary investigation of lava flows and rock formations of the Crater Lake region and, by tumbling logs over the cliffs, constructed a rude raft which took them over to Wizard Island. Representatives of a federal scientific bureau had finally discovered the remarkable lake. When Diller became chief of the Cascade Division (formerly the Division of Volcanic Geology) replacing Dutton in 1888, he did much to publicize Crater Lake among the scientific community and became a strong advocate for national park status.

The next year, 1884, Diller was assigned the task of examining the geology of Mount Shasta, while Dutton, still in charge of the geologic survey of the Cascade Range, went to New Mexico to study volcanic landforms of the southwest. 1885 would prove a significant one in the history of Crater Lake. Dutton, now the chief of the newly formed Volcanic Division of the. Geological Survey, wrote his contribution for the sixth annual report for Director Powell from Berryvale near Mount Shasta on June 30, 1885, wherein he states, "On the 18th of May [1885] I left Washington for northern California to enter upon a study of the Cascade Range, and I am at present engaged upon it."[19] Dutton headed a small military party accompanied by Joseph LeConte of the University of California. The Dutton party was to reach Crater Lake on August 15, 1885, but not before they met with and added to their expedition the man who was to figure most prominently in the creation of Crater Lake National Park, William Gladstone Steel.

William Gladstone Steel was born on September 7, 1854, in Stafford, Ohio, the son of abolitionist parents. The family moved to Pittsburgh in 1865 and then to southeastern Kansas in 1868. It was in Kansas where the young Steel read a description of the discovery of a mysterious lake in Oregon. The account inspired a young man's resolve to one day see the lake for himself, a goal which he carried with him when his family moved west again in 1872 to Portland, Oregon. Nine years passed before Steel could find anyone who could provide him with authoritative information about the lake. It was not until 1885 that he saw Crater Lake for himself.

In the summer of 1885, accompanied by a friend, J. M. Breck, Steel took the Oregon & California Railroad to Medford, where he boarded a stagecoach for Fort Klamath. At Fort Klamath, Steel and Breck met Dutton and LeConte, who were then in the midst of their Cascade investigations. This was a meeting that could only be described as fateful, for here an inspired citizen with enough fortitude and determination to lead a seventeen year crusade for national park status for Crater Lake met a leader of the scientific community. The combination of idealistic citizens exerting pressure on their elected officials, armed with petitions and arguing from positions of cultural nationalism and recreation potential, together with a scientific community vouching for the scientific value of nature's monuments was to prove decisive in the campaign to create not only Crater Lake National Park, but several other early national parks as well.

Dutton described his first impressions of the lake in an 1886 article in *Science*.

> The beauty and majesty of the scene are indescribable . . . It was touching to see the worthy but untutored people, who had ridden a hundred miles in freight-wagons to behold it, vainly striving to keep back tears as they poured forth their exclamations of wonder and joy akin to pain. Nor was it less so to see so cultivated and learned a man as my companion [LeConte? Steel?] hardly able to command himself to speak with his customary calmness.[20]

As Steel viewed Crater Lake for the first time on August 15, 1885, he was overcome by a conviction that the area should be saved in its natural state forever. For many hours in camp, Dutton, Steel, LeConte, and Breck discussed plans to preserve the area from private exploitation. All agreed that the initial task was to have the federal government withdraw the area from private settlement and development. Steel wrote, "My first visit to Crater Lake was in 1885, at which time the thought was suggested by Capt. C. E. Dutton, of having the lake and environs drawn from the market. Promptly acting on the suggestion, my friend Hon. Binger Hermann [Congressman for Oregon] was sought and a movement started looking to the formation a national park."[21]

Returning to Portland later that summer, Steel and Breck began a petition drive in the state. Steel also wrote a descriptive article on Crater Lake which he sent to some thousand or so newspapers throughout the United States asking the editors for their support in his drive for a national park. Statewide petitions were signed by Oregon's leading political, business, religious, and civic figures including the Governor Zenas E. Moody. Dutton assisted the cause by writing an article emphasizing the geologic magnificence of Crater Lake.[22] Petitions were presented to President Grover Cleveland on December 21, 1885, which asked for the withdrawal of five townships from entry and that the lands be kept and reserved as a public park. Resolutions and memorials supporting national park status for Crater Lake were forwarded to Congress by the Oregon State Legislature, Portland Board of Trade, Portland City Council and various other town and county councils throughout Oregon.

Steel traveled to Washington, D.C., himself and met with Secretary of the Interior Lucius Q. C. Lamar and President Cleveland in January 1886. His compelling arguments and petitions moved Secretary Lamar to support a recommendation written by Assistant Secretary of the Interior G. A. Jenks to withdraw from public entry the lands surrounding Crater Lake. The recommendation was forwarded to the president who issued an executive order to that effect on February 1, 1886. The executive order was made temporary in light of pending national park legislation. Oregon Senator Joseph Dolph had earlier introduced a bill on January 18, 1886, providing for the establishment of a national park. Congressmen Binger Hermann, who had helped Steel on his lobbying tour of official Washington which led to Cleveland's executive order, introduced a similar bill in the House on February 1, 1886. While under consideration, the Congress received more memorials and resolutions from organizations and newspapers from around the country. In spite of this support, opposition by timber, mining, and grazing interests to the bills was overwhelming and the supporting testimony was never reported.

As long as Crater Lake national park bills were submitted and considered in the Congress, the executive order of February 1, 1886, would stand. For every year until 1894 bills were introduced, but each time, except for some success in the Senate, they were defeated in the House, facing uncompromising opposition by timber, mining, and grazing interests. To counter their objections, Steel wrote a book, privately printed by his brother David in Portland, entitled *The Mountains of Oregon*[23] which Steel sent gratis to President Benjamin Harrison and his cabinet and to members of Congress. The book publicized his intense interest in preserving the natural beauty of not only Crater Lake but of the region as a whole. The votes in the House Committee on Public Lands on Crater Lake legislation indicate that Steel's monograph did little to change congressional minds or effectively deal with the objections of the timber and grazing interests.

The stage on which the campaign for a Crater Lake National Park had been played was altered on September 28, 1893, when President Cleveland issued a proclamation establishing the Cascade Range Forest Reserve. The reserve extended from the California border in the south along the entire length of the Cascade Range in Oregon to the Columbia River. The five townships embracing Crater Lake, withdrawn in 1886, were included in the new reserve. Yet adequate funds and regulations for the management of the forest reserve had not been appropriated or developed, allowing ranchers and shepherds to continue their practice of grazing animals within the new reserve, sometimes reaching as far as the high alpine meadows of the Cascade crest. Unauthorized timber removal persisted as well. The region's isolation no longer guaranteed immunity from private exploitation.

Steel and others who had been battling for national park status broadened their efforts by including the new Cascade Range Forest Reserve in their preservation struggle, as grazing interests lobbied for the breakup and reduction of the reserve. To better meet this new challenge, Steel founded a mountaineering club on July 17, 1894, which was formally established two days later on the summit of Mount Hood. Called the Mazamas after an endangered species of mountain goat, qualifications for membership called for an individual, male or female, to climb a Cascade summit that was perpetually covered in snow. As time went on, Steel increasingly used the Mazamas as a vehicle to assert the preservation goals of the forest reserve against what he saw as the predations of timber companies and sheepherders.

In 1896 Steel organized an outing by the Mazamas to Crater Lake in an effort to dramatize and promote conservation issues. Understanding the need for support from the federal scientific community in future conservation measures in Congress, Steel invited a number of professionals from the federal scientific establishment to join the Mazamas on their outing to Crater Lake that summer. These scientists included C. Hart Merriam, chief of the United States Biological Survey; Joseph Diller, United States Geological Survey geologist; Frederick V. Colville, chief botanist of the United States Department of Agriculture; and Barton W. Everman of the United States Fish

Commission. With their presence, the 1896 outing took on the look of a scientific expedition as they and other naturalists provided guided nature walks and campfire lectures on the flora, fauna, and geology of the region to some fifty members of the Mazamas and a few hundred other visitors. Barton W. Everman conducted research for the Fish Commission investigating water temperature and quality of the lake while Joseph Diller continued his examination of the geology of the region. The outing culminated in the christening of Crater Lake's ancestral mountain "Mount Mazama" with appropriate ceremonies on August 21, 1896.

The Mazama's 1896 outing was reported in *Harpers Weekly*,[24] and in the same year, Joseph Diller wrote an article supporting preservation of the scenic and geologic wonders of Crater Lake which appeared in the *Annual Report of the Smithsonian Institution*[25] and in the *National Geographic Magazine*[26] in 1897. But the most important result of the outing, besides the increased national awareness of Crater Lake, and no doubt one of the objectives of William Steel, was that each of the scientists participating in the 1896 outing would eventually submit written testimony in support of Crater Lake National Park bills soon to be placed before Congress.

Efforts to establish a national park were renewed after a four year hiatus by Thomas H. Tongue in 1898 when he entered the House of Representatives as the sole Congressman for Oregon. Tongue introduced national park legislation on January 25, 1898, and again on December 8, 1899, after the first bill died during debate on the House floor. But the 1899 bill also expired on the House floor, while similar legislation introduced by Oregon's two senators were never reported out of the Senate Committee on Public Lands. In both houses of Congress overwhelming opposition from grazing and timber interests and the issue of prohibiting mineral exploration in the proposed park once again spelled defeat.

The bill that would ultimately be enacted into law establishing Crater Lake National Park was introduced by Representative Tongue on December 10, 1901. The 1901 bill was virtually identical to Tongue's earlier bills. Knowing that an extra effort would be needed to pass the bill after two recent failures, Steel began circulating a petition addressed to the Senate and House of Representatives requesting favorable consideration of the pending Crater Lake legislation. By March 1902 the petition had been signed by over 4,000 people and included many leading citizens from the state of Oregon. The House bill was reported out of committee on March 11, 1902. The committee report recommended passage without amendment and included supporting letters from Binger Hermann, former representative for Oregon and in 1902 the commissioner of the General Land Office; Thomas Ryan, former acting secretary of the Interior; and Diller, Merriam, Everman and Colville, federal scientists who had joined the Mazama outing of 1896. The support of President Theodore Roosevelt was obtained enabling the bill to be brought to the full House for consideration after a delaying tactic was made by the House Speaker. The Crater Lake park bill finally passed in the House on April 21, 1902, with one significant amendment: the restriction against mining in the reservation was eliminated. The Senate passed the bill on May 9 and on May 22, 1902, President Roosevelt signed the measure into law creating the nation's sixth national park.[27] While mineral exploration never became an issue for superintendents of Crater Lake National Park and was later prohibited in all park properties in the legislation creating the National Park Service in 1916, the Congress, by permitting mineral exploration in Crater Lake National Park in 1902, indicated it had not fully come to terms with the contradiction between absolute preservation and preservation for efficient use or utilitarian notions of conservation.

Crater Lake National Park shared several features of early national parks. Like Mount Rainier, the park was restricted to the monument itself, focused within a single rectangle of land. Ecosystem integrity was not the driving force behind the early national parks; rather, it was the spirit of monumentalism and ideals of cultural nationalism which animated the early national park movement.[28] Over time, the boundaries of Crater Lake National Park were modified to include important wildlife habitats essential to the stability of the park's ecology and to promote better management. The combination of determined citizens effectively led aided by the unfailing support of the scientific community, proved to be the making of the Crater Lake National Park and other early national parks as well. The added incentive of Steel's petition of 1901-1902 and the well timed intervention of President Theodore Roosevelt provided the extra boost needed to overcome the opposition to the Crater Lake park bill. National park status for Crater Lake was the culmination of a seventeen year campaign by William Gladstone Steel, a perseverance that surprised even Oregon's congressional delegation. He was forever known after 1902 as the "Father of Crater Lake National Park" and became its second superintendent serving from July 1, 1913, to November 20, 1916.

UNITED STATES GEOLOGICAL SURVEY MAPPING ACTIVITIES, 1886-1988

While in Washington, D.C., lobbying for a national park in January to February 1886, William Steel met with John Wesley Powell, director of the United States Geological Survey and persuaded him to make a thorough survey and sounding of Crater Lake. Since the area was then being considered for national park status, Director Powell agreed, assigning Clarence E. Dutton to direct summer field work in the area and delegating to Steel the tasks of preparing the boats and gathering the necessary sounding tackle.

Steel arrived in Ashland, Oregon, by rail on July 4, 1886. Earlier, Clarence Dutton, an Army Captain of Ordnance on leave to the. Geological Survey, requested an escort from Vancouver barracks to perform the hard labor of lowering the boats and equipment from the rim down to the lake. Steel met Dutton in Ashland with ten enlisted men commanded by Captain George W. Davis. A four-oared boat of twenty-six feet and two skiffs, recently constructed in Portland, were unloaded from the train and placed on wagons. On July 7, the party left Ashland and proceeded northward along the Rogue River road (formerly the Jacksonville-Fort Klamath road) to the lake, a journey that took four days. Lowering the boat, later named the "Cleetwood," down nine hundred and fifty feet of loose stones and talus to the water's edge took several hours of hard labor but the prospect of sounding, exploring the entire shoreline, and visiting Wizard Island surely made the difficult work much easier to bear. After rowing to Wizard Island, Dutton and Steel were astonished to find that on a preliminary sounding, the weight plummeted to a depth of 1,480 feet. In all, 168 soundings were made. Dutton reported that the deepest cast measured two thousand feet. The lake's shoreline was thoroughly explored and many prominent features named, including Dutton Cliff, proposed by Steel and accepted by the geologist. Dutton examined the geology of the area and concluded that Crater Lake was formed in the same manner as other great calderas of the world, including Volcan de Taal in the Philippines, the large calderas of Central America, and those he had visited earlier in Hawaii.[29] Completing work at Crater Lake on August 5, 1886, Dutton continued on a northward march with his escort, crossing and recrossing the Cascade Range examining the geology of the mountains in detail. Steel remained at the lake to help with the topographic work. In his report of the 1886 field season, Dutton was everywhere intrigued by what he found, but upon his return to Washington, D.C., in the fall, his studies of the Cascade Range would abruptly end as he was immediately assigned to lead the investigation of the Charleston, South Carolina, earthquake of August 31, 1886. His able assistant, Joseph Silas Diller, who first saw Crater Lake in 1883, competently continued Dutton's investigation of the Cascades.

In preparation for his 1886 examination of Crater Lake, Dutton arranged to have the survey team that had completed topographic work on the six one degree by one degree, 1:250,000-scale quadrangles in the northern California Cascades move north into Oregon. Dutton explained,

> Before leaving Ashland arrangement had been made with Mr. M. B. Kerr, topographer, and his assistant Mr. Eugene Ricksecker, to meet me at the lake in the later part of July for the purpose of making a good map of the lake and of the country immediately surrounding it . . . They arrived duly, and in the course of a week they obtained the data for making a map, which has since been drawn by Mr. Kerr.[30]

This map was most probably a manuscript map made for Dutton's use on the spot and was later employed in the compilation of the *Ashland Sheet* 1:250,000 topographic quadrangle, the first map to show Crater Lake in detail.

The topographic team of Mark B. Kerr and Eugene Ricksecker surveyed the area in Oregon between the 42nd and 43rd parallels and 121st and the 123rd meridians during the field seasons of 1886 and 1887. The results were published as the first two Geological Survey topographic quadrangles for the state of Oregon, namely the *Klamath Sheet*, first published in 1889, and the *Ashland Sheet*, first published as the "Edition of Oct. 1891," both one degree by one degree, 1:250,000-scale quadrangles, contiguous and conforming to the *Modoc Lava-Bed Sheet* and *Shasta Sheet*, respectively, to the south below the 42nd parallel in California. Even though the original eastern boundary of Crater Lake National Park ran along the 122nd meridian, or along the grid line separating the *Klamath* and *Ashland* quadrangles, both maps were important to the future park, as the its boundaries were later extended in places east of the 122nd meridian. The *Ashland Sheet* showed the majority of the park's land area, but not all. The northernmost four minutes above the 43rd parallel were not included. A glance at the General Land Office's state maps of Oregon from the mid-1890s indicates a rapid absorption of the new, more detailed Geological Survey data on Crater Lake and the southern Oregon Cascades into the Land Office's maps.

These 1:250,000-scale reconnaissance maps of northern California and southern Oregon remained the only topographic maps of the region well into the 1930s. Portions of the *Ashland*, *Klamath*, *Shasta*, and *Modoc Lava-Bed* sheets were combined to produce the special Geological Survey irrigation map entitled *Klamath Project, California-Oregon: General Progress Map* published in cooperation with the United States Reclamation Service at a scale of 1:250,000 in April 1905. The *Ashland* and *Klamath* sheets were published as plates 71 and 72 respectively showing density of standing timber in the Geological Survey's *Annual Report* for 1899/1900.[31] Both the 1905 irrigation map and the forest inventory maps of 1900 showed Crater Lake and its surrounding region.

T. Brook White, the official historian for the Mazamas, wrote in his 1896 report that three maps had been received by the club: a detailed map of Mount Hood made by the General Land Office; the 1:250,000-scale *Ashland Sheet*; and a special Crater Lake map produced by the

Geological Survey, ". . . the special Crater Lake sheet being published at our [the Mazamas] request made through our representatives in Congress, but for which we also recognize our indebtedness to Mr. [Joseph Silas] Diller."[32] The *Crater Lake Special Map, Oregon*, "Edition of Mar. 1896," (fig. 10) was the direct result of the Mazamas' request through its congressional delegation to Director of the United States Geological Survey, Charles D. Walcott, and executed with the assistance of Joseph Diller, chief of the Cascade Division. The special Crater Lake map conformed to the proposed park boundaries outlined in congressional bills. After the *Klamath* and *Ashland* sheets, the special map of Crater Lake was the third Geological Survey map published for Oregon. The special map was published at a scale of 1:62,500 with a contour interval of one hundred feet and was not simply an enlargement of the Ashland and Klamath 1:250,000-scale sheets, as it continued four minutes north above the 43rd parallel. Lake soundings determined on Dutton's 1886 visit were noted in feet and along the lower margin of the map extended a diagrammatic cross section of Crater Lake drawn by Joseph Diller. A landmark map in the history of Crater Lake, it indicates the existence of an earlier preliminary map, probably the one compiled by Kerr and Ricksecker on the spot in 1886, with a contour interval of one hundred feet extending north above the 43rd parallel grid line because the map states "Surveyed in 1886" and carries no other compilation or revision date. This map was soon republished as the "Edition of Oct. 1899" in volume two of Henry Gannett's folio *Topographic Atlas of the United States, Physiographic Types* to illustrate an example of, unfortunately, "A Crater," not a caldera, with a descriptive text appearing opposite the map.[33]

Still titled *Crater Lake Special Map, Oregon*, a new edition was issued separately as the "Edition of May 1903," and to improve this landmark map, Joseph Silas Diller added a descriptive text and photographs on the back (dorso) entitled "Illustrations and Description of Crater Lake." This 1903 map of Crater Lake was the first separately published national park map produced by the Geological Survey to include a descriptive text. The same year, 1903, also saw the publication of Diller's important geological study of Crater Lake as number

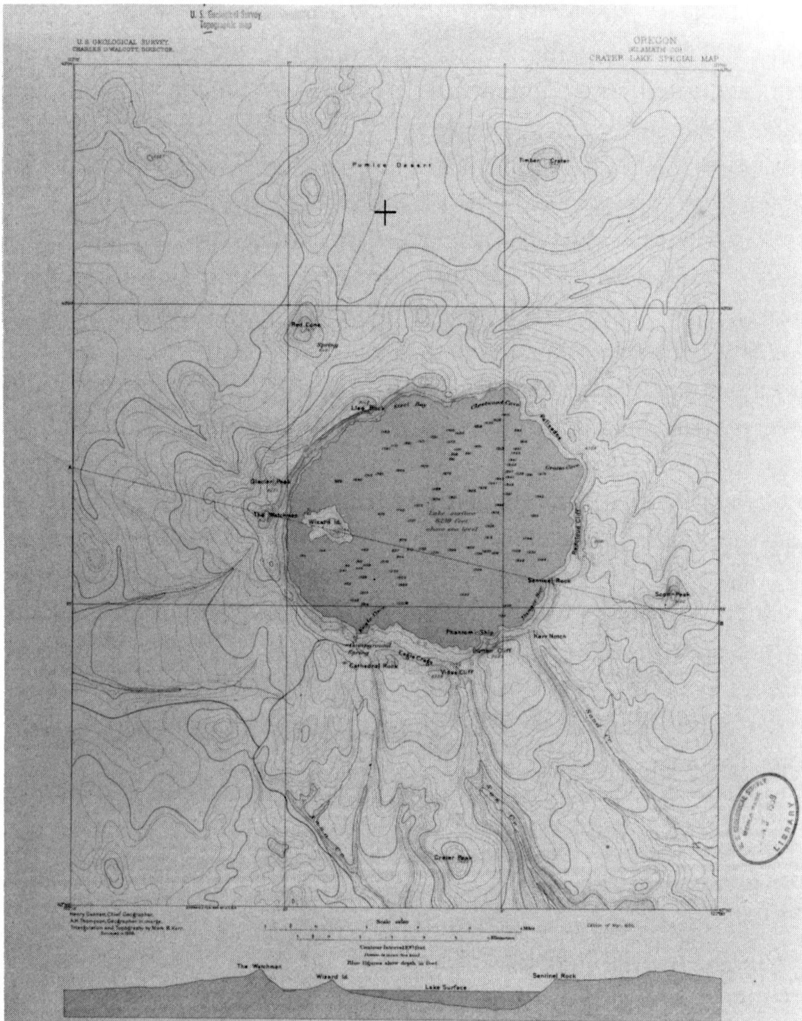
OREGON
CRATER LAKE SPECIAL MAP
Pumice Desert
Timber Crater
Red Cone
Llao Rock
Glacier Peak
The Watchman
Wizard Is.
Sentinel Rock
Scott Peak
Phantom Ship
Kerr Notch
Cathedral Rock
Vidae Cliff
Crater Peak
Wizard Id.
Lake Surface
SECTION FROM A to B

Fig. 10.

three in the Geological Survey's new flagship Professional Paper series. *The Geology and Petrography of Crater Lake National Park* contained a geologic map with a scale of 1:142,560, a map at the same scale showing locations of rock specimens described, a small-scale sketch map marking travel routes to the park, and a number of black and white photographs.[34] In the twenty years since Diller first climbed upon its rim, Crater Lake had passed from being a local attraction, imperfectly understood and mapped, to become one of only a few congressionally designated national parks whose geologic past had been deciphered and had been accurately mapped at a detailed scale.

The 1903 map was reprinted in 1910 but carried a cautionary note: "This map is out of date: revised sheet in course of publication." At the request of the Department of the Interior, the Geological Survey began a complete revision of the Crater Lake special map in 1908. A preliminary "Administrative map" was published in 1910 at scale of 1:48,000 with a contour interval of fifty feet. This map had a public land grid of township and section lines to be used as reference points for park improvements. In 1911, the map *Crater Lake National Park* was published in both an administrative edition (public land grid with a symbol key in the margin) which was not distributed to the public and an edition for general distribution. Both maps had a contour interval of fifty feet, a great improvement over the one hundred foot interval of 1903, and were far more accurate than the earlier special sheet. The edition of 1911 intended for the public had a revised descriptive text on the verso by Diller. The Geological Survey reprinted this edition in 1913 and again in 1925. The edition of 1911 was revised and reprinted in 1931 and included for the first time the road network of the park including the rim road, completed in 1919.

The Geological Survey maps of Crater Lake National Park published in the twentieth century were so far superior to all others in terms of scale and accuracy that maps published by other federal and state agencies and by the commercial sector pale by comparison. The National Park was shown on United States Forest Service and General Land Office, later Bureau of Land Management, maps in smaller scales than the 1:62,500-scale used on the Geological Survey map. The

National Park Service published park brochures that included only a small scale maps. These brochures carried a great deal of text describing the park, its rules and regulations, and other tourist information. It was not until the 1980s that the National Park Service published a truly excellent park map.

Ever since the publication of the 1896 special map of Crater Lake, the Geological Survey continued to improve its mapping of the park. A truly outstanding set of seven national park maps, including one for Crater Lake, were drawn by John H. Renshawe during the years 1914 and 1915 as part of the federal government's effort to increase internal tourism.[35] Renshawe was appointed to the Geological Survey's original topographic staff on July 8, 1879, and previous to this appointment, he served John Wesley Powell's United States Geological and Geographical Survey of the Rocky Mountains from 1873-1879. All seven of the national park maps drawn by Renshawe were engraved and printed by the Geological Survey but were distributed through the United States Government Printing Office. *Panoramic View of the Crater Lake National Park* was published in 1914 with a scale of 1:62,500, the same scale as the topographic map. The seven Renshawe national park maps are important as they represent one of the earliest attempts by the Geological Survey to depict landforms by relief shading. Not only are they pleasing to the eye, they convey the idea of physical relief much more immediately to the average map reader than the conventional topographic map and thus better fulfilled their purpose of popularizing the national parks.

The survey revised and enlarged the Crater Lake National Park map between 1931 and 1945 and published a completely new map in 1946. Francois E. Matthes, a survey geologist, contributed an expanded text for the verso. A further revision was made in 1956 which was issued in both a contour and a shaded relief edition. A new text by University of California geologist Howell Williams was added to the verso on all printings of the 1956 map after 1965 replacing the text by Matthes. Crater Lake National Park was never mapped as a conventional fifteen-minute topographic map. As part of its effort to complete 1:24,000-scale map coverage of the United States by 1991, the Geological Survey mapped and then published twenty provisional edition, 7.5-minute topographic maps covering the entire park in

1985. Continuing its tradition of excellence in the publication of special topographic maps of the national parks, the Geological Survey issued a completely new *Crater Lake National Park and Vicinity* topographic map on water resistant paper in 1988. The folding map has a color photo of the lake taken from Castle Crest on its title panel and the Howell Williams text on the verso was revised by Survey geologist Charles R. Bacon in 1984. The Geological Survey has continued to map and to examine the geology of the park down to the present day. Indeed, a large-scale geologic map of Crater Lake is currently being compiled under the direction of Charles Bacon.

The fact that an Indian war first brought Crater Lake onto a printed map seems to be in keeping with its violent past. While the lake was by-passed by all explorers and federal scientists up to 1883, once it was discovered by the United States Geological Survey it took only twenty years for the area to be accurately mapped and its geological secrets revealed. Place names such as Dutton Cliff and Kerr Notch remind us of the early involvement of the Geological Survey with the exploration of mapping of Crater Lake. However, Joseph Silas Diller, the geologist who was the first to see Crater Lake on behalf of the survey, the one who helped William Gladstone Steel to achieve park status for the area, who published the first monograph on its geology, and who helped the public to understand the region's past through his contributed text to the verso of the park maps of 1903 and 1911, sadly is missing from the many personal place names within the park. Llaos, Hillman, Steel, Applegate, Victor, Merriam, Williams, and even the boat, "Cleetwood," are remembered through place names, but not Diller. Even though a prominent peak inside Lassen Volcanic National Park has been named for Diller in connection with his geologic investigations there, a place name dedicated to his memory within the boundaries of Crater Lake National Park would be a fitting monument to the man and his contributions.

Exploration within the national park has not ceased in modern times but continues, most recently beneath the blue waters of the lake. In 1986 an exploratory geothermal well was bored just four and a half miles outside the park by the California Energy Company, a private firm. At 1,350 feet below the ground, temperatures measured 226 degrees Fahrenheit. But the United States National Park Service became concerned about the drilling. By tapping hot water deep in the earth, could the proposed geothermal energy facility affect Crater Lake's source of water or its famous clarity? Drilling was halted outside the park in order to find answers to these questions. A submersible named "Deep Rover" was brought in to explore the lake floor in search of geothermal springs in 1987 and 1988. While hot springs have been located on the bottom of America's deepest lake, their geologic relationship with test wells outside the park is still being investigated. The discovery of bacterial mats on the bottom has, however, intrigued oceanographers and plans are being made to examine further these unusual biological communities with the submersible.

Out of the destruction of Mount Mazama came a stunningly beautiful blue lake and what was once the tightly held secret of the native Americans became a national park within fifty years of its discovery by a group of wayward prospectors. The prospectors did not find gold, but they indeed found treasures in the deep blue clarity of its waters and in the purity of the air that allows the visitor to scan the horizon for more than a hundred miles.

NOTES

*This paper is based on a presentation given at the American Library Association Annual Conference, Dallas, Texas, June 1989.

1. William Gladstone Steel, *The Mountains of Oregon* (Portland, Ore.: David Steel Publishers, 1890), 12.
2. Ibid., 15-16.
3. Ella E. Clark, *Indian Legends of the Pacific Northwest* (Berkeley: University of California Press, 1953), 53-55.
4. Ruth Kirk, *Exploring Crater Lake Country* (Seattle: University of Washington Press, 1975), 23-24.
5. John Charles Fremont, *The Expeditions of John Charles Fremont.* Vol. 1, *Travels from 1838 to 1844* (Urbana: University of Illinois Press, 1970), 585.

6. Ibid. 588.
7. Carl I. Wheat, *Mapping the Transmississippi West* (San Francisco: Institute of Historical Cartography, 1957-1963), Vol. 4, 85.
8. Carl I. Wheat, "Mapping the American West, 1540-1857," *Proceedings of the American Antiquarian Society* 64 (April 1954): 165.
9. John Wesley Hillman, "Discovery of Crater Lake," *Steel Points* vol. 1, no. 2 (January 1907): 79.
10. Ibid., 79.
11. [John W. Sessions], "Headwaters of the Rogue River. Blue Lake," *Oregon Sentinel* (Jacksonville), 8 November 1862, p. 2, col. 3.
12. Ibid.
13. "Trip to Crater Lake," *Oregon Sentinel* (Jacksonville), 28 August 1869, p. 2, col. 3.
14. S. A. Clarke, "Klamath Land," *Overland Monthly* vol. 11, no. 6 (December 1873), 548-554; Francis Fuller Victor, *Atlantis Arisen, or, Talks of a Tourist About Oregon and Washington* (Philadelphia: J. P. Lippincott, 1891).
15. Israel Cook Russell, "A Geological Reconnaissance in Southern Oregon." In: *Fourth Annual Report of the United States Geological Survey, 1882-1883*, 465-473.
16. Clarence E. Dutton, *Tertiary History of the Grand Canyon District* (Washington, D.C.: U.S. Government Printing Office, 1882. U.S. Geological Survey Monograph No. 2.
17. Clarence E. Dutton, "Hawaiian Volcanoes." In: *Fourth Annual Report of the Director of the U.S. Geological Survey, 1882-1883* (Washington, D.C.: U.S. Government Printing Office, 1884), 75-219.
18. Clarence E. Dutton, "Report of C. E. Dutton, U.S. Geological Survey." In: *Fifth Annual Report of the Director of the U.S. Geological Survey, 1883-1884* (Washington, D.C.: U.S. Government Printing Office, 1885), 42-43.

19. Clarence E. Dutton, "Report of Capt. C. E. Dutton, U.S. Geological Survey, Volcanic Division." In: *Sixth Annual Report of the United States Geological Survey, 1884-1885* (Washington, D.C.: U.S. Government Printing Office, 1885), 61-62.

20. Clarence E. Dutton, "Crater Lake, Oregon, A Proposed National Park," *Science* vol. 7, no. 160 (February 26, 1886): 179.

21. William Gladstone Steel, *The Mountains of Oregon* (Portland, Ore.: David Steel Publishers, 1890), 17.

22. Clarence E. Dutton, "Crater Lake, Oregon, a Proposed National Park," *Science* vol. 7, no. 160 (February 26, 1886): 179-182.

23. William Gladstone Steel, *The Mountains of Oregon* (Portland: David Steel Publishers, 1890).

24. George W. Kirkman, "Crater Lake," *Harpers Weekly*, 19 September 1896, 932.

25. Joseph Silas Diller, "Crater Lake, Oregon." In: *Annual Report of the Board of Regents of the Smithsonian Institution*, July 1897 (Washington, D.C.: U.S. Government Printing Office, 189.7), 369-379.

26. Joseph Silas Diller, "Crater Lake, Oregon," *National Geographic Magazine* vol. 8, no. 2 (February 1897): 33-48.

27. Yellowstone, Yosemite, General Grant, Sequoia, and Mount Rainier being the others established before Crater Lake National Park.

28. Alfred Runte, *National Parks: The American Experience*, 2nd rev. ed. (Lincoln: University of Nebraska Press, 1987), 67-68.

29. Clarence E. Dutton, "Report of Capt. C. E. Dutton, U.S. Geological Survey, Division of Volcanic Geology." In: *Eight Annual Report of the United States Geological Survey, 1886-1887* (Washington, D.C. ; U.S. Government Printing Office, 1889), 158.

30. Ibid.

31. J. B. Leiburg, "Cascade Range Forest Reserve, Oregon." In: *Twenty-First Annual Report of the United States Geological Survey,*

1899-1900, Part 5, Forest Reserves and Atlas (Washington: U.S. Government Printing Office, 1900). 209-498.

32. T. Brook White, "Historians Report for 1896," *Mazama* vol. 1, no. 2:272.

33. Henry Gannett, *Topographic Atlas of the United States, Physiographic Types* vol. 2 (Washington, D.C.: U.S. Geological Survey, 1900).

34. Joseph Silas Diller, "The Geology and Petrography of Crater Lake National Park," U.S. Geological Survey Professional Paper No. 3 (Washington, D.C.: U.S. Government Printing Office, 1903).

35. The seven national parks drawn by Renshawe were Crater Lake, Glacier, Mesa Verde, Mount Rainier, Rocky Mountain, Yellowstone, and Yosemite.

SELECTED CARTOBIBLIOGRAPHY

Allan Cartography. *Crater Lake & Mount Mazama* [map]. 1:62,500. Medford, Oregon: Raven Maps & Images, 1989.

Habersham, Robert A. *Habersham's Sectional and County Map of Oregon Compiled From the Most Recent U.S. Government Maps and Surveys* [map]. Scale of statute miles 12 to 1 inch. Portland, Ore.: J. K. Gill & Co. ; New York: engraved and printed by G. W. & C. B. Colton & Co., 1874.

Renshawe, John H. *Panoramic View of the Crater Lake National Park, Oregon* [map]. 1:62,500. Washington, D.C.: U.S. Government Printing Office ; engraved and printed by the U.S. Geological Survey, [1914].

Symons, Thomas W. *Map of the Department of the Columbia* [map]. [1:1,000,000]. Vancouver Barracks, Washington: Engineer Office, Department of the Columbia, 1881.

U.S. Forest Service. *Crater National Forest, Oregon* [map]. [1:510,000]. [Washington, D.C.]: U.S. Forest Service ; printed by the U.S. Geological Survey, [1917].

U.S. General Land Office. *Diagram of Oregon* [map]. 18 miles to 1". [Washington, D.C.]: General Land Office, 1866.

U.S. General Land Office. *A Diagram of Public Surveys in Oregon* [map]. 18 miles to 1". [Washington, D.C.]: General Land Office, 1863.

U.S. General Land Office. *State of Oregon, 1876* [map]. 15 miles to 1". Washington, D.C.: General Land Office ; New York: Photo lith & print. by Julius Bien, 1876.

U.S. Geological Survey. *Administrative Map of Crater Lake National Park* [map]. 1:48,000. Washington, D.C.: USGS, 1909.

U.S. Geological Survey. *Ashland Sheet, Oregon* [map]. 1:250,000. Washington, D.C.: USGS, 1891.

U.S. Geological Survey. *Crater Lake National Park (Klamath County), Oregon* [map]. 1:62,500. Washington, D.C.: USGS, 1911. Reprinted 1913 and 1925. Reprinted with corrections, 1931.

U.S. Geological Survey. *Crater Lake National Park and Vicinity, Oregon* [map]. 1:62,500. Washington, D.C.: USGS, 1946. Reprinted 1957.

U.S. Geological Survey. *Crater Lake National Park and Vicinity, Oregon* [map]. 1:62,500. Washington, D.C.: USGS, 1956, printed 1959. Contour and shaded relief editions.

U.S. Geological Survey. *Crater Lake National Park and Vicinity, Oregon* [map]. 1:62,500. Reston, Va.: USGS, 1988.

U.S. Geological Survey. *Crater Lake Special Map, Oregon* [map] 1:62,500. Washington, D.C.: USGS, 1896.

U.S. Geological Survey. *Crater Lake Special Map, Oregon* [map] 1:62,500. Washington, D.C.: USGS, 1899. In: Gannett, Henry. *Topographic Atlas of the United States, Physiographic Types.* Washington, D.C.: USGS, 1899.

U.S. Geological Survey. *Crater Lake Special Map, Oregon* [map]. 1:62,500. Washington, D.C.: USGS, 1903.

U.S. Geological Survey. *Klamath Project, California-Oregon, General Progress Map [map].* 1:250,000. Map 6092. Washington, D.C.: U.S. Geological Survey and U.S. Reclamation Service, 1905.

U.S. Geological Survey. *Klamath Sheet, Oregon* [map]. 1:250,000. Washington, D.C.: USGS, 1889.

U.S. Geological Survey. *Land Classification and Density of Standing Timber: Ashland Quadrangle, Oregon* [map]. 1:250,000. Washington, D.C.: USGS, [1900].

U.S. Geological Survey. *Land Classification and Density of Standing Timber: Klamath Quadrangle, Oregon* [map]. 1:250,000. Washington, D.C.: USGS, [1900].

U.S. National Park Service. *Automobile Guide Map Showing Roads in the Crater Lake National Park, Oregon* [map]. [1:63,360]. Washington, D.C.: U.S. National Park Service, 1922.

Warren, G. K. *Map of the Territory of the United States from the Mississippi River to the Pacific Ocean . . . to accompany the Reports for a Railroad Route* [map]. 1:3,000,000. Washington, D.C.: War Department, 1857.

Williamson, R. S. and H. L. Abbot. . . . *Routes in Oregon and California, Map No. 2, From the Northern Boundary of California to the Columbia River* [map]. 1:760,320. Washington, D.C.: War Department, 1855.

". . . THE MOST SUBLIME OF ALL EARTHLY SPECTACLES." EXPLORATION AND MAPPING OF THE GRAND CANYON

Charles A. Seavey

School of Library Science
University of Arizona

The Grand Canyon is in an extremely remote area of the United States and was virtually unknown prior to 1860. Joseph Christmas Ives, the first Anglo explorer to publish a report on his visit to the canyon used the term "famous" in 1859, yet there had been only two previous expeditions to the area and the accounts of neither had been published in English.[1] The exploration and mapping of the Grand Canyon of the Colorado River in Arizona is more than just a fascinating story in its own right.[2] There are several facets to the story. One is a straight historical narrative; another facet encompasses how knowledge about the canyon spread. Yet another facet of the story is the proliferation and effect of the maps published by various United States government explorers in the mid-nineteenth century. The first question to be addressed is Ives' use of the word "famous." What was the extent of knowledge about the canyon prior to 1858?

PRE-1858 KNOWLEDGE: THE WRITTEN RECORD AND THE ORAL TRADITION

Pueblo Indians lived in the canyon for about four generations in the thirteenth century, and Garcia Lopez de Cardenas, a captain in the service of Vasques de Coronado, visited there in 1540 looking for the seven cities of Cibola. Although there is some debate about Cardenas' exact location, Smith places him at Desert View, within the present national park[3]. Cardenas himself notes "a river the banks of which were so high that they seemed to be three or four leagues in the air." The report stated that "some rocks, which appeared from above to be the height of a man, were higher than the tower of the Cathedral of Seville."[4] This was the first recorded European sight of the canyon.

The report of the Coronado expedition, including a section on the Cardenas journey, was recorded in manuscript form by Pedro de Castaneda as *Relacion de la jornada de Cibola conpuesta por Pedro de Castaneda de Nacera*... in 1596. The first published (printed) version of the *Relacion* did not appear until 1838 when it was translated into French by Henri Ternaux-Compans and published for the first time by A. Bertrand, Paris, 1838. The translation is based on the original manuscript copy in the Lenox Branch (collection) of the New York

Public Library.[5] The first version in English did not appear until the historian George Parker Winship published his translation and the original Spanish text in 1896.

Although the 1838 printed version of the *Relacion* does not appear in the catalog of the War Department Library, it is reasonable to assume that a copy was available to the United States Topographical Engineers (USTE), the United States' exploration agency prior to the Civil War.[6] Ives, or someone in the Topographical Engineers, could read French as Ives' report contains a brief summary of Cardenas' account of his time at the canyon. Ives did not cite his source, simply referring to the "history" in his summary. Nothing in Ives' summary differs from Winship's translation of 1896. Thus, while an account of the canyon had been published prior to the 1850s it was not widely accessible and does not appear to have been a major factor in spreading knowledge of the canyon beyond scientific circles.

After the Cardenas expedition the canyon dropped out of western consciousness for 236 years. In 1776 two Spanish priests, Francisco Dominguez and Silvestre de Escalante, visited the area. According to Ted J. Warner, in the latest scholarship on the Dominguez-Escalante Expedition, the two priests got to within approximately 18 miles of the North Rim at roughly 113° west and then skirted the canyon where the Colorado runs northeast at 111°30' west, crossing the river at the present Arizona-Utah state line.[7]

Warner notes that the original manuscript diary kept by Dominguez and Esclante is no longer extant…or at least nobody knows where it is. There are at least nine manuscript copies of the original diary. The 1976 translation, by Fray Angelico Chavez, is based on the copy in the Edward E. Ayers Collection at the Newberry Library in Chicago which was the first copy made from the original in Santa Fe in early 1777. Another copy, made in July of 1777, is now in the Archivo General De Indias, Seville. The next copy is dated 1792 and is in the Archivo General de la Nacion (Mexico City). Other manuscript copies are extant, including one in the Library of Congress which appears to be the only one, other than the Newberry copy, in the United States. The Library of Congress copy was obtained as part of the Peter Force Collection in 1867 and hence was unavailable to Ives in the late 1850s.[8]

The first printed version was produced in Mexico, in Spanish, in 1854 as part of *Documentos par la Historia de Mexico.* Neither Warner nor Fray Chavez regard the 1854 version as reliable or accurate. Subsequent English translations appeared in 1909,[9] 1945,[10] and 1951.[11] All three of the earlier English translations relied upon the 1854 printed version, with occasional spot checks against the Newberry manuscript.

John Wesley Powell, the first man to bring the canyon to widespread popular attention, had apparently either read Escalante's journal in Spanish (printed or manuscript) or may have seen earlier manuscript translations of which we are as yet unaware. In any case, Powell devoted a lengthy paragraph to the Dominguez-Escalante expedition with particular attention to where they crossed the Colorado on their return to Santa Fe, "from the description which we have read we were enabled to determine the place."[12]

The available published accounts of previous expeditions to the Grand Canyon available to the Topographical Engineers at the time of the Ives expedition were in French or Spanish and published outside the United States. Based on the available evidence, Ives' observation about the "famous" canyon cannot have been based on extensive written records. The knowledge must have been based on the oral tradition.

THE ORAL TRADITION

In the early nineteenth century trappers and mountain men came to the Southwest, and some saw the canyon. Smith makes it clear that there were a number of trappers, traders, and adventurers roaming the Southwest from approximately 1800 until the period under consideration.[13] Because few of the mountain men left records (let alone made maps) there is still debate about where they were and exactly what they saw. There is one known instance of a map made by a fur trapper, Warren Augustus Ferris, in 1836 that clearly shows the Grand Canyon.[14] Unfortunately the map was not published until 1940! Ferris did publish a series of articles in the *Western Literary Messenger* in the 1840's, including one titled "Chanion of the Colorado" in

1842.[15] Although Ives refers to the stories of "one or two trappers," it is apparent from his account that he had not read Ferris' article and may have considerably underestimated the number of trappers and mountain men who had seen the canyon.[16]

The Ferris map shows the Grand Canyon extending as far north as Coeur d'Alene, Idaho. The map compresses the area it depicts so exact location of the area labeled "Great Chanion of the Colorado" is difficult to determine.[17] The labeled area may well be currently known as Glen Canyon, to the north of the present national park. Exact location aside, the point remains that contemporary evidence indicates that the approximate location of the canyon was known by the mountain men well prior to the Ives expedition. There was also knowledge to be gained from native Americans.

Writing about the canyon, Powell relates that "enough had been seen to foment rumor, and many wonderful stories have been told in the hunter's cabin and prospector's camp." He further relates a native American myth about the creation of the canyon and states that he had been warned by the Indians not to enter the canyon as they considered it disobedience to the gods.[18] Powell was not without a flair for the dramatic, but it is unlikely there is much exaggeration in his account of knowledge of the canyon in the oral traditions of both the mountain man and the native American. In any case, at least some knowledge of the magnificent canyon filtered back to the East, although there are no contemporary maps extant from the period, since the Ferris map did not see light until 1940. Ives' reference to the "famous" canyon is probably justified if we accept the idea that its fame was limited to those familiar with the West and western exploration.

UNITED STATES GOVERNMENT INTEREST, 1850s

Transmission of knowledge about the Grand Canyon depended largely upon the government explorer with scientific training who went, saw, and *reported in reproducible form* what had been seen. In this sense, the mapping of the Grand Canyon may be seen as a model for the mapping of the entire western United States. There is a progression of government explorers at the Grand Canyon who represent all save the very first stages of United States government exploration of the American West. The first stage is represented by explorers like Lewis and Clark or Zebulon Pike, government employees working more or less independently of any specific government agency in their explorations. The second stage is the explorations of the United States Topographical Engineers (USTE), an organized branch of the Army which operated in the American West from approximately 1840 until the Civil War. The third stage of government exploration, a transitional one, lasted from the end of the Civil War until 1879; during this time government exploration made the transition from a military operation to one controlled by civilians. Four separate agencies of the government, some civilian, some military, were responsible for mapping various portions of the American West. The final stage of government mapping started in 1879 when the formation of the purely civilian United States Geological Survey (USGS) consolidated all western exploration into one agency.

The final three stages of government exploration are all represented in the maps of the Grand Canyon. The maps produced during these periods were widely reproduced and disseminated knowledge of the canyon in two ways, through maps appearing in various publications of the United States Government Printing Office and by influencing commercial map makers.

Located in what was one of the most inaccessible areas of North America, the canyon was visited by few until the late nineteenth century. There are great cartographic questions about any location: Where is it? What is it shaped like? How does it relate to its general area? These questions about the Grand Canyon were basically answered between 1859, when the first map to receive widespread reproduction and distribution was made, and 1906 when the first Geological Survey large scale topographic quadrangle map of part of the canyon was produced. Other maps since 1906, which improve on original interpretations or present different sorts of knowledge about the canyon, must also be considered.

THE FIRST AMERICAN EXPEDITION: IVES, 1857-1858

Although its existence was certainly known, the Grand Canyon was not by itself the real objective of the first scientific expeditions. It was simply as part of the Colorado River, usually referred to as Colorado River of the West, that the exploration of the Grand Canyon began.

At the time of these early expeditions, communication in the West was rudimentary at best. The transcontinental railroad would not be completed until 1869. Overland travel was slow, tedious, fraught with danger from Mother Nature and uncertain relations with native Americans. Since the discovery of America communication has largely followed the path of least resistance: water. The St. Lawrence, the Great Lakes, the Ohio, the Mississippi, and myriads of lesser waterways provided the channels along which the young nation expanded. Waterway communication served in the West, too. The Platte and the Missouri system provided some access on the northern plains, and in the Southwest the Colorado, a short overland haul or sea voyage from southern California, was an obvious entryway into otherwise difficult terrain.

In 1856 the California Legislature petitioned Congress to have the Colorado explored from (Fort) Yuma (present day Yuma, Arizona) to its source.[19] California's interest dovetailed with Washington's, and in December 1857 Lt. Joseph Christmas Ives, United States Topographical Engineers (USTE), and his party arrived at the mouth of the river with the purpose of discovering the navigational possibilities of the largely unknown stream.[20]

Although the primary mission of the expedition was to discover how far the river was navigable for steamboats, this was no straightforward military reconnaissance. In keeping with more or less standard USTE procedures of the time, the expedition had scientific interests as well. John Strong Newberry was appointed physician, with the added responsibility of investigating natural history and geology. Heinrich Mollhausen, representing the great German geographer Alexander von Humboldt, was the expedition's collector of natural history. F. W. von Egloffstein was topographer with two other men making astronomical and meteorological observations. Egloffstein had previously been with a Fremont expedition, and he, Newberry, and Mollhausen had all been involved in the Pacific Railroad surveys.

In keeping with the main purpose, their means of transportation was an iron steamboat, 50 feet long. Built in sections in Philadelphia, it was taken apart and shipped via Panama to San Francisco, where the expedition gathered, and thence to the mouth of the Colorado to be assembled prior to the ascent. Arriving at Robinson's Landing on the 28th of November, 1857, the expedition started putting its craft together, and shortly after midnight on December 21st the "Explorer" headed north.

By early March they had reached an area roughly 20 miles below the current site of Hoover Dam. At this point the "Explorer" struck a submerged rock, dislodging the boiler and generally creating havoc. After three days of repairs, Ives determined to go no farther without a reconnaissance via their rowboat. Ives and his companions rowed upriver through the Black Canyon to a point slightly north of the present location of Hoover Dam. The river was full of rapids, and Ives decided that the limits of practical navigation had been reached and that he would drop downriver, divide the party, and make an eastward overland reconnaissance to connect the river and the Mormon wagon route to San Bernadino.

Ives returned downriver to a point just north of the junction of the California, Nevada, and Arizona borders in what is now the Camp Mohave Indian Reservation. All the notes and collections gathered to that point were sent south on the "Explorer." Ives, Newberry, Mollhausen, Egloffstein, assorted assistants, and twenty troopers started overland on the 23rd of March, 1858, heading generally northeast, making scientific observations along the way. A sort of scientific nadir was hit at this stage of the journey. The party had hired some Indian guides, and others had wandered in and out of contact with the expedition. One morning a Hualpais turned up, apparently in search of food. Ives describes him as having "features like a toad's and the most villainous countenance I ever saw on a human being." Mollhausen, in a burst of scientific overzealousness, suggested that they should "take" the poor Indian and pickle him in alcohol as a zoological specimen! Cooler heads prevailed, and the Hualpais went his way.[21]

In early April the party started its descent into the canyon. The terrain was such that Ives had to draw an unworldly comparison:

> The place grew wilder and grander . . . before long we were hemmed in by walls two thousand feet high . . . the increasing magnitude of the colossal piles that blocked the end of the vista and the corresponding depth and gloom of the gaping chasms into which we were plunging imparted an unearthly character to a way that might have resembled the portals of the infernal regions. Harsh screams issuing from aerial recesses in the canyon sides and apparitions of goblin-like figures perched in the rifts and hollows gave an odd reality to this impression. The idea suggested itself that were the guides to desert us, our experience might further resemble that of the dwellers in the unblest abodes. . .in the difficulty of getting out.[22]

Ives and party became the first non-natives to walk the floor of the Grand Canyon. It was a "sublime moment in the history of American Exploration." [23]

Ives' foreboding about getting back out of the canyon nearly came true. They spent two days wandering about after the guides deserted but finally emerged and established a base camp. Ives and Egloffstein, with a dozen men, started further exploration. After several frustrating days, which included having to haul Egloffstein, whose curiosity got the better of him, back up a precipice by making a rope of the troopers' carbine slings, the attempt to return to the bottom was abandoned.

The expedition moved south and then east approximately along the line of current Interstate 40, crossed the Little Colorado on May 2nd, and reached Fort Defiance (Arizona) on May 23rd, 1858. The expedition broke up at this point, most of the party continuing east. Ives went by stage to El Paso and thence back to Fort Yuma where he gathered the collections and field notes from the early portion of the expedition. After selling the "Explorer" he proceeded to California to take a steamer back to Washington.

There were two important aspects of Ives' *Report*, published in 1861. First was the simple fact of having been there. Ives' descriptions, and Egloffstein and Mollhausen's drawings, conveyed for the first time to the scientific community and the general public the enormity and spectacle of the canyon. Ives was wrong in his statement that this would "doubtless be the last party of whites to visit this profitless locality" for his very report simply whetted curiosity.[24]

The other important aspect of the report was Egloffstein's maps. Two maps were published with the report in 1861. "Map 1" is at a scale of 1:380,160, in four sections each measuring approximately 14.5 by 8.5 inches, and covers the area from the Gulf of California north along the Colorado to Fortification Rock, approximately the current site of Hoover Dam. "Map 2" (fig. 11) is at a scale of 1:760,320 and measures approximately 14.75 by 35 inches. It covers an area from roughly 34°30' north (the level of Prescott, Arizona) to just north of the Utah border, and from 108°30' west (just east of the current Arizona-New Mexico line) to 115°30' west (just west of present Las Vegas, Nevada). "Map 2" recapitulates most of section 3 and all of section 4 from "Map 1" and includes all of the eastern overland portion of Ives' journey; "Map 2" is the first map to attempt to convey the enormity of the canyon in visual terms.

In order to do this, Egloffstein abandoned hachure marks, the conventional method of the time for indicating elevation changes, and made the maps using a variation of shaded relief technique.[25] Shaded relief had been used before, but Egloffstein refined it in order to show different elevations of level terrain. Working directly from field notes he engraved shapes and outlines onto steel plates. He then cut very fine parallel lines over the whole plate. Inspection of even the white areas on the map with a magnifying glass shows fine lines .005 of an inch apart. Each area was then exposed to acid for varying lengths of time, the higher elevations receiving the least, consequently making the lightest impression on the printed version. The lower elevations received more acid, thereby producing a darker image on the finished map. The effect is very similar to that produced by today's familiar shaded relief method and was well suited to depicting the impression the canyon creates.

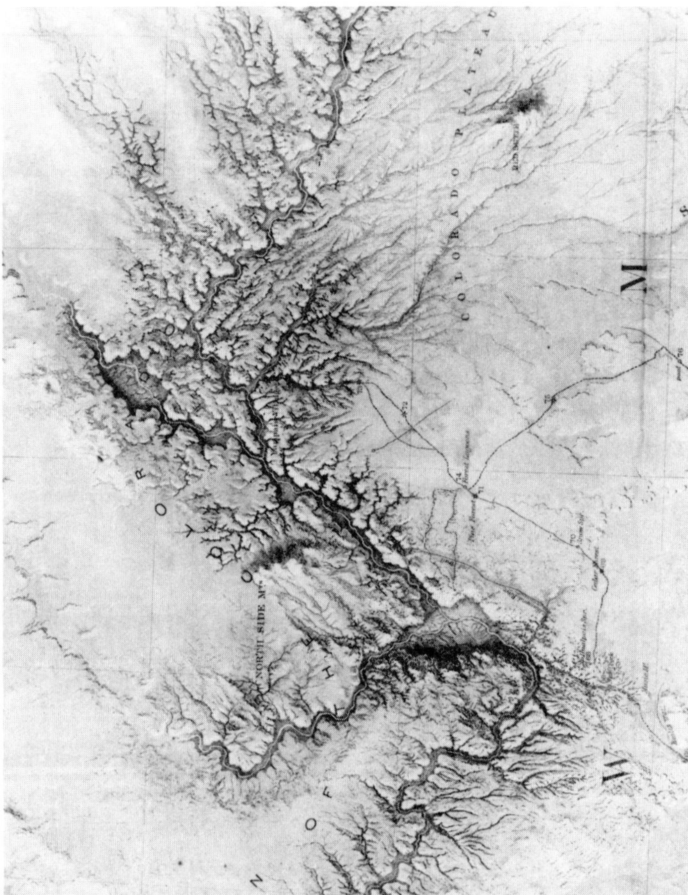

Fig. 11. Detail from *Map no. 2: Rio Colorado of the West . . .* by Joseph Christmas Ives, 1861. (Courtesy of Special Collections, University of Arizona Library)

The technique is a tricky one, although it produces beautiful maps. Unfortunately it was not a method that could easily be applied to lithography, then starting to come into use as a major map printing process. The only other notable map done in this method was one by Egloffstein for the Macomb expedition of 1859.

If the technique produced a lovely map, Ives' interpretation of his data produced a major error of location that lasted well into the 1870s. Ives' error was in mislocating the junction of the Colorado and Little Colorado rivers by approximately 90 miles. As Ives did not label the confluence of the two rivers on the map, there is a certain amount of ambiguity as to what junction he thought was the proper one. Because Ives' interpretation had a considerable impact upon commercial (and government) maps of the area, the question deserves fuller attention.

In order to eliminate the possibility that Ives was simply making gross errors in deriving latitude and longitude from his astronomical observations, several of his locations were plotted on modern maps. Camp 67 is located at the "Big Cañon, mouth of Diamond River." Diamond Creek is the name still employed for a stream running into the Colorado at that point, so there is a precisely located spot with which to compare Ives' observations. Ives' plot for the camp is about five miles west of his actual location. Other locations were spot checked and the five mile east-west error was the maximum observed. This hardly accounts for the mis-location of the confluence of the two rivers. It appears then that Ives simply misjudged another confluence as the one in question. There are two confluences shown on the map, and there are two possible explanations for each as to what Ives' interpretation was.

As Ives' map shows, the eastern portion of the river running east-west is labeled the "Little Colorado or Flax River." The western portion of the river is labeled "Colorado River." The confluence clearly has to be between the two labels. The first possibility for the junction of the Little Colorado and Colorado is north of Camp 67 in the middle of the large chasm depicted on the map. That chasm lies at roughly 113° 20' west...where the Colorado is running almost due north in what is now the Hualapai Indian Reservation. Two rivers join there. One, unnamed, heads northwesterly from the course of the main river shown on the map. This junction will be referred to as "Junction 1." The second junction is just above the area labeled "Yavapai Villages" at approximately 113° west. At this junction an unnamed river flows from the north-northeast into the main river. The Ives map was compared to modern 1:250,000 and 1:24,000 Geological Survey maps of the areas in question. There are two possibilities for Junction 1. Parashant Canyon cuts northwest from the main stream approximately 13 miles north of Diamond Creek; Trail Canyon cuts northwest approximately five miles north of Diamond Creek. The distance from Diamond Creek to the confluence on Ives' map is approximately 10 miles. Parashant Canyon is much the larger of the two and appears to be the more likely candidate for the junction depicted. The modern maps do not show a steady stream disgorging from either Parashant or Trail Canyons, but Ives was there in the spring, and the intermittent streams may well have been full at the time.

Although, as shall be shown, some commercial map makers adopted Junction 1 as the confluence of the two rivers, it is clear from the *text*, if not the map, that Junction 2, just above "Yavapai Villages," was the junction that Ives had in mind.

After establishing Camp 73 and making a side trip down the Cataract Creek-Havasu Canyon complex, the party started retracing their path south towards Camp 71. They discovered a side trail leading towards the main canyon. Ives and a small party took the side trail and headed north till dusk forced a stop.

> The country became rough and so much cut up by ravines that it was impossible to approach very closely to the main river. A good view was obtained of the walls of the Flax (i.e., Little Colorado) river cañon, and its *mouth approximately located* (emphasis added). The junction was below the mouth of Cascade Creek, showing that that stream is not, as had been supposed, a tributary of the Colorado, but of its smaller affluent.[26]

As Ives was describing this, he was looking north and to the west (assuming "below" to mean downstream) of the point where Cascade Creek joins the Colorado. The problem is that "Cascade Creek" is no longer used to identify any stream in this area. The name does not appear in the Geological Survey's *Geographic Names Information System* printout for Arizona. *Arizona Place Names* uses the name, citing Ives as the authority, and places the creek within the Cataract Creek-Havasu Canyon complex.[27] This interpretation correlates with Ives' text. Ives appears to have used the term Cascade Creek for what we now identify as Havasu Creek. In the quote above it is clear that Ives' Cascade Creek runs directly into the Colorado. The only stream that meets that criteria that fits with Ives' description of where he was is the present Havasu Creek.

There are two possible areas that are both west of Havasu-Cascade Creek that come in from the northerly direction taken by Ives' "Colorado." Both Toroweap Valley and Tuckup Canyon meet the criteria. Tuckup Canyon is approximately 6.5 miles west of Havasu Creek, and eleven miles east of Toroweap Valley. Neither the Toroweap Valley, nor Tuckup Canyon support a full-time stream joining the Colorado.

Given that we cannot be certain of Ives' location when he "approximately located" the confluence he thought was that of the Colorado and Little Colorado, other than that he was 10-12 miles south, and in uncertain light, the evidence is simply insufficient to make a judgement as to whether Toroweap Valley or Tuckup Canyon is the more likely candidate to contain the supposed Colorado River. The trend of Ives' "Little Colorado" is generally close to that followed by the actual Colorado River as far east as approximately 112° west, roughly the site of the current park headquarters and tourist area. From 112° eastward however, Ives departs from the actual course. The Colorado swings north towards Utah at approximately 111°50' west, the mouth of the Little Colorado being located approximately 12 miles north of the bend. The Little Colorado flows generally from the southeast, and is properly located on the eastern portions of Ives' map. Ives had misjudged the confluence, and since the eastern part of the trip was well south of the actual river, and he could not make direct observations, he had interpolated the single stream between the two points he positively located.

The ambiguity of Ives' map as to the confluence of the Colorado and Little Colorado is reflected in the maps of both commercial and government cartographers who based their maps on Ives. At least one commercial publisher, on *Hartley's Map of Arizona from Official Documents,* judged that Junction 1 was the confluence.

In general, however, both commercial and government cartographers read Ives' text as well as looking at the map, and used Junction 2 as the confluence of the two rivers. The General Land Office *Sketch of Public Surveys in New Mexico and Arizona for 1866* uses Junction 2, as does Richard Gird's 1865 *Official Map of the Territory of Arizona, With All the Recent Explorations.* The distinguished map publishing firm of Samuel Mitchell adopted Junction 2 for its *County Map of Arizona and New Mexico* of 1872. Mitchell, in fact, used Junction 2 on the same map as late as 1877, long after more accurate data (the Wheeler *Atlas* sheets) were available.

Ives' *Report upon the Colorado River of the West* was published in 1861, at which time the divided nation had larger problems on its mind. Ives joined the Confederacy, exploration came to a halt, and the Topographical Engineers were absorbed into the Engineer Corps.

THE TRANSITIONAL PERIOD, 1865-1879: POWELL AND WHEELER

Although government exploration in the Southwest ceased with the start of the Civil War, others still turned up there. Wheeler reports finding, in 1871, a monument left by one O. D. Gass in 1864. James White, a prospector, claims to have descended through the canyon on a raft in 1867, although the authenticity of this story is still a subject of debate. With the war over, and things more or less back to normal, eyes once more turned westward. One of the most remarkable men in the history of the West was about to meet its greatest river. John Wesley Powell and the Colorado would make each other famous.

Powell's story is so familiar that it does not need detailed exposition, but some recapitulation is necessary. As a major in the United States Army he was wounded at Shiloh, lost his right arm, and eventually left the service. He taught natural science and geology at a number of colleges in Illinois and started gathering financial support for his western adventures. After a preliminary trip into Colorado,

Powell and nine men set out, on May 24th, 1869, from Green River, Utah, on their epic descent of the river.[28]

By August 13, 1869, the party was ready "to start on our way down the Great Unknown." Making only a few miles a day, they remained in the canyon for 16 days. Food was scarce and not of the best quality. The river was fierce, there were accidents, and by August 26th Powell's scientific detachment was so far strained that he wrote, "A few days like this and we are out of prison."[29] A few days later they were out. "They had done what no other men had done before. They had dared the great river and they had won, and in doing so they dispelled the last great mystery of the American West . . . Powell returned to civilization a hero."[30]

He came back to the Colorado in 1870, and again in 1871-1872. By this time he had gotten financial support from Congress by lining up recommendations from the Interior Department, the War Department, and the Smithsonian Institution. After 1874 Powell headed the United States Geographical and Geological surveys of the Rocky Mountain Region and his explorations became an official arm of the government.

A book, rather than a map, was Powell's immediate contribution to the knowledge of the canyon: *Exploration of the Colorado River of the West.* As a scientific work, it reshaped some of the geological thinking about the formation of the canyon. As an adventurous travelogue, it was a huge success, selling out immediately after it rolled off the presses at the Government Printing Office.[31] Powell, never adverse to the possibilities of publicity, published a less technical, abridged version of *Exploration* in a series of articles in *Scribner's Monthly Magazine* in 1875.[32]

Although one inevitably associates the canyon with Powell, the only map he produced was a relatively minor affair. Published in 1875 to accompany one of the *Scribner's Monthly Magazine* articles about the 1870 trip, the map measures only 5 7/8 by 4 1/8 inches.[33] While the course of the river is considerably more accurate than that shown on the Ives/Egloffstein map, it lacks that map's elegance in portraying the enormous contrasts of relief in the canyon. The triangulation and topography done by the Powell survey, however, had a considerable impact on the maps produced by the Geological Survey after 1879.

GEORGE M. WHEELER: LAST STAND OF THE ARMY EXPLORER

While Powell was in the limelight, a considerably less well known endeavor was also in the field, and while not as well publicized as Powell's successes, it made large contributions towards the understanding and mapping of the canyon. George M. Wheeler, a young Army lieutenant (West Point, 1866), became head of the last purely military exploratory agency. Assigned as an engineer on the staff of the commanding general of the Department of California, he was ordered to explore, among other things, ". . . to the head of navigation on the Colorado River."[34]

Wheeler's first expedition was an 1869 exploration into the deep southwest that produced much needed information about an area of which precious little was known. His rise was not entirely due to his good work, however. The Army establishment was mindful of the monopoly the regulars had enjoyed in the halcyon days prior to the Civil War when the Topographical Engineers had been *the* exploratory arm of the government. In contrast, the work of Powell and F. V. Hayden was sponsored by the Interior Department, and Clarence King, while nominally under the aegis of the War Department, ran an almost purely civilian operation. The fact that Brigadier General A. A. Humphreys, head of the Corps of Engineers, had been a charter member of the Topographical Engineers in 1838 added impetus to have a purely military survey in the field.

Wheeler took full advantage of the opportunity. His expedition of 1871 was a spectacular journey that started in California, moved eastward, traversed Death Valley, and then ascended the Colorado River into the Grand Canyon. Wheeler reasoned that far more information could be gathered by moving slowly upstream, than by traveling rapidly downstream as Powell was even then doing.

The party, which would travel in three specially constructed rowboats and a barge, was the usual mix of scientists and support types including a photographer, a naturalist, two topographers, a barometric assistant, and "general assistant" F. W. Loring, who was also a reporter for an eastern newspaper.[35] Most importantly, Grove Karl Gilbert, the great geologist, was along for his first extensive western trip. It was from this expedition, and that of 1872, that Gilbert would form the notion of ancient Lake Bonneville.[36]

Leaving Camp Mohave (in the present Camp Mohave Indian Reservation) just north of the California-Nevada line, the party started upstream on September 16, 1871, perhaps wishing that Ives' old "Explorer" were there to provide the motive power. By September 25th they were passing Fortification Rock, where Ives had been stopped 12 years before. Turning eastward they rowed upstream towards a pre-planned meeting with a relief column at Diamond Creek.

On October 9th they passed the point where O. D. Gass had turned back in 1864. Wheeler noted that "the rapids are more formidable than any yet seen." Two days later one of the boats was wrecked in a rapid, destroying a large part of Wheeler's notes, the astronomical and meteorological observations, surveying instruments, and "worse, all the entire rations of that boat." Wheeler admitted to being "much dispirited." By October 16th things were bad enough that Wheeler named that day's stopping point "Starvation Camp." Two days later he dispatched two of his men to go ahead and find the relief expedition as quickly as possible. The main party reached Diamond Creek the next day, and the relief column came up on the 20th.

What had Wheeler accomplished? He had successfully completed an expedition that combined the best elements of drama and scientific reporting. It established him as a "name" to be reckoned with in the field of scientific exploration. That name, and the Army's desire to re-establish some portion of their previous hegemony over western exploration, led in 1872 to the creation of the United States Geographical Surveys West of the 100th Meridian as an official agency of the War Department. It was to be the "Last Stand of the Army Explorer."[37]

Cartographically speaking, Wheeler made a major contribution towards accurate depiction of the Grand Canyon. Part of the publication program of the Wheeler survey involved a series of maps that collectively are known as the Wheeler "Atlas."[38] Sheet 67 covers the area of the present Grand Canyon National Park while sheet 66 covers the area west of sheet 67, well into California and Nevada. Both maps cover 2°45' longitude by 1°40' latitude at a scale of 1:506,880, measure roughly 14.5 by 17.5 inches, and show relief by hachure marks. Sheets 66 and 67 correct Ives' erroneous interpretation of the junction of Colorado and Little Colorado Rivers. Wheeler's maps are essentially accurate in their location of the Colorado River and in the details of the actual course of the river. This is in contrast to the later maps produced by the Geological Survey based on work by the Powell survey.

Wheeler also produced, in his report on the 1871 expedition, a map covering the western half of the canyon and farther west to the California line.[39] The map shows the itinerary of part of the 1871 expedition and includes fairly detailed topography. Wheat characterizes it as ". . . a document valuable in the record of Western exploration."[40]

In fact Wheeler's 1871 trip was the last of the spectaculars of exploration and survival in the canyon. Exploration of the area became systematized in Wheeler's survey, and Powell's. Both agencies would have parties in the field doing the unromantic, but necessary, work of laying out baselines and establishing astronomical stations, making the trigonometrical surveys necessary to producing accurate large scale maps.

In conjunction with the Geological and Geographical Survey of the Territories (Hayden) and the Geological Exploration of the Fortieth Parallel (King), the Powell and Wheeler surveys represented the transitional stage of government mapping in the West. The four surveys represent both a continuation and a departure from the work of the Topographical Engineers. They were a continuation in that their basic mission was an orderly and scientific government directed exploration and that their reports, like those of earlier government explorers, would be published by the government. The departure from previous practice is that the surveys were a transition between the purely military operations that preceded them and the purely civilian Geological Survey which took their place. Hayden and

Powell operated under the aegis of the Interior Department; King was a civilian, but reported to the Chief of Engineers of the Army; Wheeler ran the only purely military operation in the group.

The four surveys pursued their various ends for a decade. There was a certain amount of infighting over funding, and touches of personal acrimony among the four directors, but generally each went its own way. In 1874 the first suggestion that the surveys be consolidated were put forth, but nothing came of it. In 1878 Congress asked the National Academy of Sciences to consider the situation. The result was a report recommending, among other things, the creation of the United States Geological Survey (USGS) in the Interior Department and the discontinuation of the four surveys. After some sharp political maneuvering, the necessary legislation was passed and signed in 1879 (20 Stat. 394) and the USGS became the nation's primary mapping agency.

USGS AND THE CANYON: 1879-1906

The next major event in the mapping of the canyon was the publication, in 1882, of Clarence E. Dutton's monumental *Tertiary History of the Grand Canyon District.*[41] It was the first major publication of the Geological Survey. When USGS superseded the King, Powell, Hayden, and Wheeler surveys it took over all their work and much of their personnel. Dutton and his *History* are an example of this. Work on the *History* had begun while Dutton was still working for the Powell survey. William H. Holmes, "perhaps the greatest artist-topographer and man of many talents that the West ever produced," who did the stunning panoramas for the accompanying *Atlas,* was an alumnus of the Hayden survey.[42] The first director of USGS, under whom the book was published, was Clarence King who had run his own survey.

The *History* was a culmination of all previous work on the canyon and a beginning upon which later geologists would build. Like G. K. Warren's great 1857 map of the West, all that remained was to "fill in the details." In all, maps at three scales cover the canyon area. The first, Plate III, "Sketch Map of the Western Part of the Plateau Province . . . ," is at a scale of 1:1,000,000. This map, although crude in some ways and employing hachure marks, is relatively accurate as to placement of the Colorado River and location of the Grand Canyon. Two maps at 1:250,000 (Plates XXII and XXIII) show the generalized geology of the Canyon District. Plate XXIII states that "this region has been well surveyed topographically, but has been reconnoitred only by the geologists." This statement is difficult to reconcile with the rather large errors in location and shape that the map contains. Comparison with the modern 1:250,000 sheet shows that the older maps have the course of the river offset some three miles to the south, and there is a pronounced southerly bend that dips below the 36th parallel just east of Bright Angel that does not really exist.

Four sheets (Plates XI, XII, XIII, and XIV) show the geology of the Grand Canyon itself from the junction of the Colorado and Little Colorado to approximately the western boundary of the present national park. Although drawn at a large enough scale, 1:63,360, that one could expect a good degree of accuracy, these sheets suffer from the same defects that the 1:250,000 sheets do. Plate XIV shows the Colorado running very close to the 36th parallel and dipping south of it at the non-existent bend east of Bright Angel. This does not coincide with the picture presented by the 1:1,000,000 map, or even with Powell's little *Scribners Monthly* map.

A possible explanation is that these maps were simply intended as rough background for the geological data being presented. This is inconsistent with both the statement on Plate XXIII and the amount of topographic detail shown on the maps themselves. A more likely, though less palatable, explanation is that the triangulation and topography done by the Powell survey was not as accurate as it might have been. This suggestion is supported by a close look at the next important step in mapping the canyon.

The first standard series of topographic maps to cover the canyon was at the 1:250,000 scale. Two sheets of the 1° x 1° "Reconnaissance Map" series cover the national park area: *Kaibab*, published by USGS in 1886, and *Echo Cliffs*, published in 1891. Although published by USGS, the triangulation and topography had largely been done by the Powell survey.[43] Even when superseded by the 1:48,000-scale maps, they enjoyed a long run, being reprinted as late as 1937. Like the

1:250,000 and 1:63,360 maps in Dutton's *Atlas*, they are offset to the south and have the same bend in the river dipping below the 36th parallel.

There is evidence to suggest that at least one member of the USGS was aware of the inaccuracies of the plates in the *Atlas*. Charles D. Walcott, a USGS geologist and subsequent director of the survey, was doing geologic fieldwork in the canyon in the fall of 1882. He used, as a base map, what appears to be a copy of Plate XIV (the 1:63,360 geologic map) from the *Atlas*. On the copy in the archives at the Grand Canyon the dip south of the 36th parallel is partially crossed out, and a new line of the river is penciled in.[44] While the new route is closer to being accurate than the base map, it does not address the 3-4 mile southerly offset at all. Walcott's field notes from that season make no mention of the accuracy of the map.[45] Walcott apparently never saw fit to report the discrepancy to USGS. The original copy of his map was, until its loss, in the files at the canyon. The 1:250,000 maps of 1886 and 1891 are no more accurate than the *Atlas*, nor do they carry any of the place names that Walcott had penciled in on his field map.

The effect of the publication of the southern course (as opposed to Wheeler's correct northern interpretation) for the Colorado River on commercial maps was the same as the various interpretations of Ives' river junction. Both versions of the location of the Grand Canyon appeared on commercial maps for several years. While the majority of the maps in the collection of the University of Arizona use the northern interpretation, there are a number that do not. E. A. Eckhoff and P. Riecker's 1880 *Official Map of the Territory of Arizona* and Frank A. Gray's *Arizona and New Mexico* of 1884 both use the southern course. As late as 1911, the major map publisher C. S. Hammond used the southern course on its map of Arizona.

LARGE SCALE MAPPING: FRANCOIS EMILE MATTHES, 1902-1906

The 1:250,000 quadrangles, while generally better and certainly more widely distributed than anything done previously, were still a long way from the 1:62,500-scale or degree of accuracy that USGS envisioned for the *Topographic Atlas of the United States*. Large scale mapping done with new data was called for, and USGS sent one of its best. Francois Emile Matthes, a native of Holland, had lived in Switzerland and had studied at Frankfurt and the Massachusetts Institute of Technology. He eventually would work for the Interior Department for 51 years, mapping not only the Grand Canyon, but the Yosemite Valley, the San Andreas Fault and Mount Rainier, and would write *Geologic History of the Yosemite Valley,* the standard work on the subject.[46] His account of mapping the canyon reveals a man with a great deal of zest for life and his work, a finely tuned sense of humor, and a will that no mere two hundred mile long canyon could defeat.[47]

Matthes arrived at what he calls "the glorious *forte-fortissimo* finale to the thousand mile long rhapsody of gorges and canyons of the Colorado River" in late winter of 1902. Traveling via the newly constructed railroad spur to the south rim, Matthes and party first saw the canyon on a grim day heavy with the promise of snow. "There it lay, a scene of sullen, stony, grimness, overwhelming by its vastness, baffling with its chaotic profusion of chiseled detail, but utterly devoid of charm. Without a word we severally slunk back to our tent." After two years' work mapping, however, Matthes reckoned that "it seemed a hardship to leave it for the commonplace world of man."

Matthes worked along the south rim for several months, establishing bench marks, plane table stations, and taking observations. The heat was incredible, work having to start in the early morning because by noon the air bubbles in the levels would disappear as the water heated up. Water, in fact, was a major problem as it had to be hauled in, often from Del Rio, 120 miles away. By midsummer Matthes decided that the party should move across to the north rim. The only problem was getting there. Rather than make a 300 mile detour he elected to cross the river at Mystic Spring, 25 miles west of his location near the railhead. The trail was a rugged one and the camp paraphernalia was cut to a minimum. Although there was a boat, it turned out to be on the wrong side. They had to swim the river, emerging exhausted and mud-covered on the north bank. In addition they discovered that there was no shelving shore on the south side, making transportation of the pack animals difficult. Finally, a simultaneous pull on the lead rope from the boat and a shove from behind ". . . sent the unsuspecting beasts head over heels into deep water." After recovering from the initial shock they then swam across behind the boat. The party worked the north rim until well into the fall, sometimes observing the tourists on the south rim through the surveying instrument. Fortunately they discovered that the Bright Angel Trail was now useable, and returned over that route, rather than the "ugly" Mystic Spring Trail.

The second year of work was much the same: water shortages, tired animals, difficulty crossing the river, and primitive living. But there were compensations. "Nowhere had we seen such overwhelming panoramas. Nor was there ever more paradisiacal land to dwell in." Working the Walhalla Plateau, which projects into the great bend, Matthes likened his party to gods, masters of all they surveyed! In the next sentence, however, he confesses that "one of the would-be gods . . . got ignominiously lost in his domain" while hunting.

Matthes was determined to finish the job and pressed on well into winter. The crew was well seasoned, the camping gear again reduced, and they kept working. "We had been eating from a sheet of canvas spread on the floor for the last six weeks. What a contrast with the nobby camp of the earlier months with its air-tight stoves, acetylene lamps, and dainty napkins." It was mid-January 1904 when the work was finished, the crew disbanded, and Matthes installed himself "on the soft cushions of a Pullman car" to return to Washington to translate his field work into maps.

Matthes' first map published was *Bright Angel*, dated 1903 and printed in 1906 (fig. 12).[48] It was at a scale of 1:48,000 in a 15-minute format, so in size it is very close to a modern 1:24,000, 7.5-minute quadrangle. With the publication of *Bright Angel* the main questions about the canyon, and its relationship to the surrounding topography, were settled. Further expeditions, like Claude Birdseye's in 1927, would do some tidying up, but the main job was done. The canyon and the relationship of the Colorado and Little Colorado rivers were located correctly, and Matthes' elegant contours conveyed the complex topography of the canyon far better than any previous effort.

Other quadrangles followed *Bright Angel*, and by 1927 USGS could release its first *Grand Canyon National Park* map at 1:48,000. The 1927

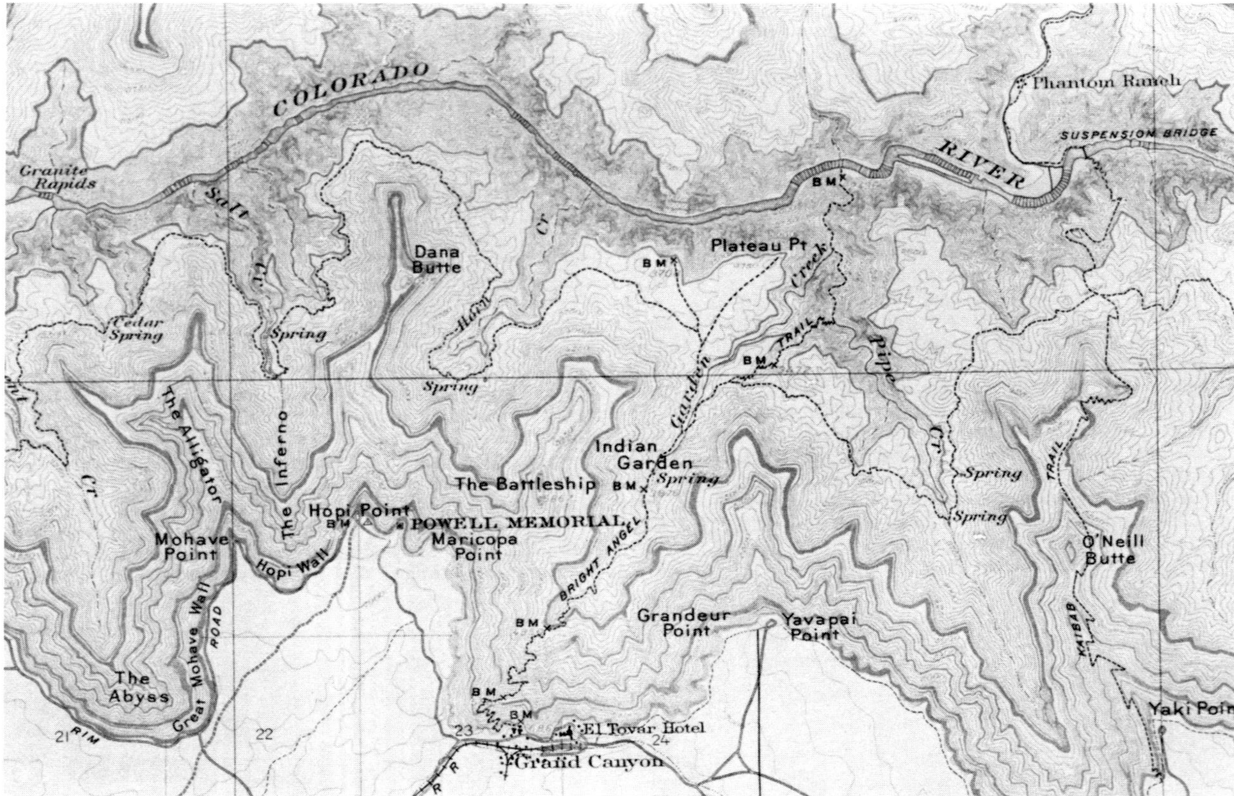

Fig. 12. Detail from Bright Angel by Frances Emile Matthes for U.S. Geological Survey, 1906. (Courtesy of University of Arizona Map Collection)

map is in two sheets and bears the unmistakable style of Matthes. Later maps include the 1:48,000 *Grand Canyon National Monument* in 1944, the monument being the area immediately west of the park. Matthes' original work served as the basis for all the later 1:48,000 special maps. *Bright Angel* itself was reprinted as late as 1961. The 1:48,000 quadrangles were, regrettably, superseded by 1:62,500 quadrangles in 1962, and in the same year USGS produced a single sheet 1:62,500 map of the park. In 1972 USGS published the beautiful shaded relief edition and in 1988 released 32 standard 1:24,000, 7.5 by 7.5-minute quadrangles covering the entire park. The great federal effort to map the Grand Canyon, starting with Ives in 1859, was finally complete.

Attempting to describe all the dozens, if not hundreds of commercially produced maps of the canyon is outside the scope of this paper. Two efforts of the commercial cartographers art, however, deserve special recognition. First is the *Geologic Map of the Grand Canyon National Park, Arizona* produced by the Grand Canyon Natural History Association in conjunction with the Museum of Northern Arizona. Using the USGS 1:62,500 *Grand Canyon National Park and Vicinity, Arizona* as a base map the *Geologic Map* . . . portrays in vivid colors the complex geology of the region. The effect, particularly when seen from a distance, is that of a work of abstract art, not a scientific document.

The second map to be mentioned is the National Geographic Society's exquisite 1:24,000 *Heart of the Grand Canyon*, a work of art in its own right. Issued in July 1978, this map surpasses both Egloffstein and Matthes in its depiction of the topographic complexity of the canyon. The National Geographic Society spared very little in the preparation of *Heart*, even importing cartographers from the Survey of Switzerland to do the cliff hachures supplementing the contour lines. The 1988 USGS quadrangles are at the same scale and are certainly an improvement over the lifeless 1:62,500 quadrangles of 1962, but even they pale in comparison to the National Geographic product.

The basic work of mapping the Grand Canyon was completed with the Matthes quadrangles starting in 1906. Commercial cartography almost always follows government surveys, and the USGS maps provide the base map for most of the trail maps and thematic maps of the canyon available at the national park bookstore, or most map collections. With the possible exception of Alaska, the Southwest presented the last territory in the United States to be accurately mapped, and the Grand Canyon provides a microcosm for the study of the progression of mapping the Southwest. It is also much more than that.

From the moment when J. C. Ives arrived at the floor of the canyon until the time that Francois Matthes settled back into his Pullman coach only a little over 44 years passed. Ives thought he would be the last white man to visit the canyon, but Matthes amused himself by watching tourists through his surveying instrument. The contrasts increase as we come forward in time. The once "impassable" Bright Angel Trail is now a virtual superhighway among trails, and one has to make reservations six months or more in advance to camp at the bottom of the canyon. The "Great Unknown" that Powell approached has now been measured and analyzed within an inch of its life. Wild adventure and heroic survival have given way to aerial photography and comfortable quarters at the Denver Federal Center. Yet not one of the men who explored and measured and mapped during those 44 years, nor anybody who has seen it since, has been unmoved by the spectacle of the canyon. Even today, beset by thousands of tourists every year and "park regulations" of great length it remains among ". . . the most sublime of all earthly spectacles."[49]

NOTES

*This paper is based on a presentation given at the American Library Association Annual Conference, New Orleans, Louisiana, June 1988.

1. Joseph Christmas Ives, *Report Upon the Colorado River of the West, Explored in 1857 and 1858.* U.S. Engineer Department. (Washington, D.C.: U.S. GPO, 1861).
2. Portions of this article originally appeared in the Special Libraries Association, Geography and Map Division *Bulletin* no. 119 (March, 1980): 4-15.

3. Melvin T. Smith. "The Colorado River: Its History in the Lower Canyons Area" (Ph.D. diss., Brigham Young University, 1972), 70.

4. George Parker Winship. *The Coronado Expedition, 1540-1542.* U.S. Bureau of American Ethnology, 14th Annual Report, part 1. 54th Cong., 2nd sess., H. doc. 230 (Serial 3531, volume 1): 329-613. (Washington, D.C.: U.S. GPO, 1896). See also OCLC 5883787 which has a mistake, stating that the manuscript used dates from 1506- clearly an error.

5. Winship, 339.

6. David Fitzgerald. *Alphabetical Catalogue of the War Department Library* . . . (Washington, D.C.: U.S. GPO, 1882).

7. Ted J. Warner, ed., *The Dominguez-Escalante Journal; Their Expedition Through Colorado, Utah, Arizona, and New Mexico in 1776.* Translated by Fray Angelico Chavez. (Provo, UT: Brigham Young University Press, 1976).

8. Library of Congress Manuscript Division letter to the author, March 29, 1988.

9. W. R. Harris, *The Catholic Church In Utah* (Salt Lake City, UT, 1909).

10. Herbert S. Auerbach, *Father Escalante's Journal, 1776-1777*(Salt Lake City, UT: Utah State Historical Society, 1943).

11. Herbert E. Bolton, *Pageant in the Wilderness: The Story of the Escalante Expedition to the Interior Basin, 1776*(Salt Lake City, UT: Utah State Historical Society, 1951).

12. John Wesley Powell, *Exploration of the Colorado River of the West and its Tributaries.* U.S. Smithsonian Institution. (Washington, D.C.: U.S. GPO, 1875), 72.

13. Smith, *The Colorado River,* 92-138.

14. See entry 416 in volume 2 of Carl I. Wheat's. *Mapping the Transmississippi West, 1540-1861.* 5 volumes (San Francisco: The Institute of Historical Cartography, 1957-1963).

15. The *Western Literary Messenger* had a "precarious existence" from 1841 until it ceased publication in 1857. See Paul C.

Phillips, ed., *Life in the Rocky Mountains . . . by W. A. Ferris* (Denver, CO: The Old West Publishing Company, 1940), pp. xxiv-xxx for details. The article, "Chanion of the Colorado" appeared in the July 20, 1842, issue of the . . . *Messenger* and is reprinted in the same book on pp. 319-320. The canyon, according to Ferris, starts at the junction of the Green and Grand rivers and stretches two to three hundred miles. Beyond a brief description of the enormous depths of the canyon, there is little detail.

16. Ives, 21.

17. The best estimate is that the north-south dimension of the area shown on the map is compressed by 130 miles, the east-west dimension by 250 miles. See Phillips, *Life in the Rocky Mountains* . . . , xii.

18. Powell, *Exploration*, 6-7.

19. U.S. Congress. *Resolution of the Legislature of California*. 34th Cong., 1st sess., Senate Miscellaneous Document 39, (Serial 835) (Washington, D.C.: U.S. GPO, 1856).

20. Ives.

21. Ives, 97-98.

22. Ibid., 100.

23. William H. Goetzmann, *Army Exploration in the American West* (New Haven, CT: Yale University Press, 1959), 390.

24. Ives, 110.

25. Ibid., appendix D.

26. Ibid, 110.

27. Will C. Barnes, *Arizona Place Names*, revised edition (Tucson, AZ: University of Arizona Press, 1965).

28. Powell, *Exploration*, 8.

29. Ibid., 96.

30. William H. Goetzmann, *Exploration and Empire: The Soldier and the Scientist in the Winning of the American West* (New York: Alfred A. Knopf, 1966), 551.

31. Ibid., 556.

32. John Wesley Powell, "The Cañons of the Colorado," *Scribner's Monthly*, vol. 9, nos. 3, 4, and 5 (January, February, and March 1875).

33. John Wesley Powell, "An Overland Trip to the Grand Cañon," *Scribner's Monthly*, 10 (October, 1875): 659-678. The map appears on page 661. It is also reproduced in Wheat, volume 5, page 354.

34. Wheat, vol. 5: 337.

35. George Montague Wheeler, *Geographical Report.* Final report (monograph), U.S. Geographical Surveys West of the 100th Meridian, vol. 1. (Washington, D.C.: U.S. GPO, 1889).

36. Goetzmann, *Exploration,* 470.

37. Ibid., chapter 12.

38. There is no actual bibliographic entity to cite here. For a description of the Wheeler Atlas see Laurence Schmeckebier's *Catalog and Index of the Publications of the Hayden, King, Powell, and Wheeler Surveys.* U.S. Geological Survey *Bulletin* 222. (Washington, D.C.: U.S. GPO, 1904): 60-65.

39. George Montague Wheeler, *Preliminary Report Concerning Explorations and Surveys Principally in Nevada and Arizona* U.S. Engineer Department. (Washington, D.C.: U.S. GPO, 1872).

40. Wheat, vol. 5, part 2: 243.

41. Clarence E. Dutton, *Tertiary History of the Grand Cañon District; with Atlas* USGS *Monograph* no. 2. (Washington, D.C.: U.S. GPO, 1882).

42. Goetzmann, *Exploration,* 512.

43. The maps themselves carry statements to that effect.

44. Charles D. Walcott, Field Map. Xerox copy located in the archives of the Grand Canyon Museum (GRCA 28090). The original is no longer extant. The map is attributed to Walcott. Much of the handwriting on the map appears to this author to be in the same hand as that in Walcott's field notes (note below).

45. Charles D. Walcott, *Field Notes, 1882*. . . . Xerox copy located in the archives of the Grand Canyon Museum (GRCA 28089).

46. The background material on Matthes is drawn from J. Donald Hughes, *The Story of Man at the Grand Canyon*, Grand Canyon Natural History Association *Bulletin* no. 14 (Flagstaff, AZ: The Association, 1967). See also the *New York Times* obituary of June 25, 1948, and the article in the *National Cyclopaedia of American Biography*. Matthes would be an excellent topic for a full length biography.

47. Unless otherwise noted the following account is drawn from Francois Emile Matthes, "Mapping the Grand Canyon," *The Technology Review* 7 (January, 1905): 1-25.

48. Illustration 12 is an excerpt from the 1906 map. The present tourist center complex and Matthes campgrounds are just east of the El Tovar Hotel at the bottom of the illustration. The parallel running horizontally through the middle of the illustration is the 36º5' N line, the smaller gridlines in the SW quadrant are part of the township and range system. As always, the illustration does not begin to do justice to the original.

49. Dutton, *Tertiary History*, 142.

SELECTED CARTOBIBLIOGRAPHY

Eckhoff, Emil A. and P. Riecker. *Official Map of the Territory of Arizona* [map]. 1:887,040. New York, NY: Graphic Company, 1880.

Gird, Richard. *Official Map of the Territory of Arizona, With All the Recent Explorations* [map]. 1" = 15 mi. San Francisco: Richard Gird, 1865.

Grand Canyon Natural History Association. *Geologic Map of the Grand Canyon National Park* [map]. 1:62,500. Flagstaff, AZ: The Association, 1976.

Gray, Frank A. *Arizona and New Mexico* [map]. 1:3,168,000. n.p.: Frank A. Gray, 1884.

C. S. Hammond & Co. *Arizona* [map]. 1:1,480,000. New York, NY: C. S. Hammond, 1911.

Hartley, William B. *Hartley's Map of Arizona from Official Documents* [map]. 1:1,570,000. New York, NY: William B. Hartley, 1863.

Ives, Joseph Christmas. *Map no. 1: Rio Colorado of the West* . . . [map]. 1:380,160. Julius Bien, Lithography. Washington, D.C.: U.S. Engineer Department, 1861. In: Ives, Joseph Christmas. *Report Upon the Colorado River of the West, Explored in 1857 and 1858.* Washington, D.C.: U.S. Government Printing Office, 1861.

Ives, Joseph Christmas. *Map no. 2: Rio Colorado of the West* . . . [map]. 1:760,320. Julius Bien, Lithography. Washington: D.C.: U.S. Engineer Department, 1861. In: Ives, Joseph Christmas. *Report Upon the Colorado River of the West, Explored in 1857 and 1858.* Washington, D.C.: U.S. Government Printing Office, 1861.

Mitchell, Samuel Augustus. *County Map of Arizona and New Mexico* [map]. 1:3,500,000. Philadelphia, PA: Samuel Augustus Mitchell, 1872.

Mitchell, Samuel Augustus. *County Map of Arizona and New Mexico* [map]. 1:3,500,000. Philadelphia, PA: Samuel Augustus Mitchell, 1877.

National Geographic Society. *Heart of the Grand Canyon, Grand Canyon National Park, Arizona* [map]. 1:24,000. Washington, D.C.: The Society, 1978.

There are actually two versions of the map which fit this description. The smaller was issued with the July 1978 issue of the *National Geographic Magazine.* The second, a sales item, is wider and extends coverage by 1.5 miles on the west side of the map and about 3,000 feet on the east side.

Powell, John Wesley. *Map of the Grand Cañon of the Colorado, Showing Route Traveled by Maj. Powell* [map]. ca. 1:2,850,000. In: Powell, John Wesley. "An Overland Trip to the Grand Cañon," *Scribner's Monthly* 10 (October, 1875): 661.

U.S. General Land Office. *Sketch of Public Surveys in New Mexico and Arizona for 1866* [map]. 1:1,800,000. New York, NY: Major and Knapp, 1866.

U.S. Geological Survey. *Bright Angel* [map]. 1:48,000. Topography series. Washington, D.C.: USGS, 1906.

U.S. Geological Survey. *Echo Cliffs* [map]. 1:250,000. Reconnaissance Map Series. n.p.: USGS, 1891.

U.S. Geological Survey. *Geologic Map of the Colorado Plateau and San Francisco Mountains* [map]. 1:250,000. Atlas plate XXIII. In: Dutton, Clarence E. *Tertiary History of the Grand Cañon District; with Atlas.* U.S. Geological Survey Monograph 2. Washington, D.C.: U.S. Government Printing Office, 1882.

U.S. Geological Survey. *Geologic Map of the Grand Cañon in the Kaibab Plateau* [map]. 1:63,360. Atlas plate XIII. In: Dutton, Clarence E. *Tertiary History of the District; with Atlas.* U.S. Geological Survey Monograph 2. Washington, D.C.: U.S. Government Printing Office, 1882.

U.S. Geological Survey. *Geologic Map of the Southern Part of the Kaibab Plateau* [map]. 1:63,360. Atlas plates XI and XII. In: Dutton, Clarence E. *Tertiary History of the Grand Cañon District; with Atlas.* U.S. Geological Survey Monograph 2. Washington, D.C.: U.S. Government Printing Office, 1882.

U.S. Geological Survey. *Geologic Map of the Southern Part of the Kaibab Plateau Head of the Grand Cañon* [map]. 1:63,360. Atlas plate XIV. In: Dutton, Clarence E. *Tertiary History of the Grand Cañon District; with Atlas.* U.S. Geological Survey Monograph 2. Washington, D.C.: U.S. Government Printing Office, 1882.

U.S. Geological Survey. *Geologic Map Showing the Kanab, Paria and Marble Cañon Platforms* [map]. 1:250,000. Atlas plate XXII. In: Dutton, Clarence E. *Tertiary History of the Grand Cañon District; with Atlas.* U.S. Geological Survey Monograph 2. Washington, D.C.: U.S. Government Printing Office, 1882.

U.S. Geological Survey. *Grand Canyon National Monument, Arizona* [map]. 1:48,000. Washington, D.C.: USGS, 1944.

U.S. Geological Survey. *Grand Canyon National Park and Vicinity, Arizona* [map]. 1:62,500. Washington, D.C.: USGS, 1962.

U.S. Geological Survey. *Grand Canyon National Park and Vicinity, Arizona* [map]. 1:62,500. shaded relief edition. Washington, D.C.: USGS, 1972.

U.S. Geological Survey. *Grand Canyon National Park, Arizona* [map]. 1:48,000. 2 sheets. Washington, D.C.: USGS, 1927.

U.S. Geological Survey. *Kaibab* [map]. 1:250,000. Reconnaissance Map Series. n.p.: USGS, 1886.

U.S. Geological Survey. *Sketch Map of the Western Part of the Plateau Province Showing the Faults of the Grand Cañon District and High Plateaus* [map]. 1:1,000,000. Atlas plate III. In: Dutton, Clarence E. *Tertiary History of the Grand Cañon District; with Atlas.* U.S. Geological Survey Monograph 2. Washington, D.C.: U.S. Government Printing Office, 1882.

Wheeler, George M. *Parts of E. California, S.E. Nevada, N.W. Arizona and S.W. Utah, Atlas Sheet No. 66* [map]. 1:506,880. Washington, D.C.: U.S. Geographical Surveys West of the One Hundredth Meridian, 1872.

Wheeler, George M. *Parts of Northern & Western Arizona & Southern Utah, Atlas Sheet No. 67* [map]. 1:506,880. Washington, D.C.: U.S. Geographical Surveys West of the One Hundredth Meridian, 1872.

Wheeler, George M. *Preliminary Topographical Map* [map]. ca. 1:1,530,000. In: George M. Wheeler. *Preliminary Reports Concerning Explorations and Surveys Principally in Nevada and Arizona.* Washington, D.C.: U.S. Geographical Surveys West of the One Hundredth Meridian, 1872.

KEEPING TRACK OF PARADISE: MAPS OF THE HAWAIIAN NATIONAL PARKS

Riley Moffat

Joseph F. Smith Library
Brigham Young University, Hawaii

Karen Stockton

Bishop Museum
Honolulu, Hawaii

Mabel Suzuki

Hamilton Library
University of Hawaii at Manoa

Hawaii's two national parks are dynamic and very unique national natural treasures; they are changing and growing in a very literal sense as America's most active volcanoes and are probably the most consistently active in the world. Rising from the sea bed 18,000 feet beneath the waves to over 13,000 feet above sea level, the mountains that made Hawaii continue to build upon themselves. Maps play an important role in studying and documenting the dynamic and ever changing nature of these parks and their environment rather than merely locating attractions.

Beginning seventy million years ago, molten lava oozed through a hot spot in the earth's crust where the island of Hawaii is now. At the same time the Pacific Plate continued to creep resolutely northwestward over the hot spot at the rate of four inches a year. When the conduit from the hot spot to the volcano bent too far, a new volcano would begin forming to the southeast over the stationary hot spot. The remnants of these ancient volcanoes now stretch northwest all the way to the tip of Alaska's Aleutian Islands.

As these volcanoes moved away from the hot spot and became dormant, the weathering forces of nature wore the dome shaped volcanoes down. This process is very evident as one moves northwest from the Big Island of Hawaii island by island. The major lavas forming the island of Hawaii are less than 700,000 years old and the major lavas of the island of Maui are no more than twice that age...veritable infants in geologic time. Even now a new undersea volcano is being formed thirty miles southeast of Hawaii. Named Loihi, it, too, may rise above the waves in a few thousand years.

The volcanoes of Hawaii are unique in that they are not only the most active but they tend to erupt "quietly." A quiet eruption is, of course, relative as opposed to an explosive or phreatic eruption such as Mt. St. Helens. Most historic Hawaiian eruptions have allowed close observation and study, making them a prime attraction for tourists and scientists. It is difficult to identify more accessible active volcanoes anywhere else in the world. Gavan Daws describes the allure of the volcanoes in his book *Hawaii: The Islands of Life*:

> In the furious world of volcanic eruptions Kilauea was as gentle and generous as a volcano could be, active almost perpetually, giving out especially liquid lavas that often fountained spectacularly, making for wonderful viewing, but not normally going off with a deadly bang. People most times ran toward Kilauea to watch, rather than away for their lives. And when the volcano was quiet, tourists could saunter down into Halemaumau with an egg in a pan and fry it on moving lava, and write home about this strange, entertaining breakfast, being sure before mailing their postal card to scorch its edges a toasty Halemaumau brown.[1]

While the desire to experience and study live volcanoes is obviously the major factor in the creation of Hawaii Volcanoes National Park and Haleakala National Park, a wide range of other factors is involved. Within the parks' boundaries are all the climatic zones from tropical to polar each with their own communities of indigenous and exotic flora and fauna.

A goal of the parks is to preserve and study the unique ecology of Hawaii. The Hawaiian Islands are young and isolated. Plants and animals had difficulty getting there and once there evolved to fit their environment. The few varieties of plants and animals that were able to reach the sterile islands of lava rock diversified to thrive in the various environments in Hawaii. Within relatively small areas an original migrant species transformed itself to adapt to desert, scrub, forest, or bog in tropical, temperate, or alpine climate zones. More than 150 kinds of natural communities can be identified in Hawaii. Man's impact in terms of changing the environment and introducing exotic new species has become more pronounced, and it is important to preserve some of the more pristine sections of the islands.

Hawaii Volcanoes National Park on the island of Hawaii and Haleakala National Park on Maui are truly the most dynamic parks in the country. Maps graphically reveal this. For example, the best evidence to date the last eruption of Haleakala is through the maps of early explorers La Perouse and Vancouver. On the south coast of Maui, La Perouse's 1786 chart shows a broad shallow bay between Puu Olai and Cape Hanamanioa, whereas Vancouver's chart made in 1793 reveals the prominent peninsula between these two points that appears on current charts. In this case, maps provide concrete evidence to corroborate oral tradition that Haleakala's last eruption occurred about 1790 low on its southwest flank.

On the island of Hawaii change continues. In the past century and a half, billions of cubic yards of lava have poured out of the various conduits from the hot spot deep under the Big Island. The extent and date of these terrestrial additions has been well documented on maps for over a century. While these eruptions enlarge the island there is often earthquake activity and the tsunamis or tidal waves they create.

This awesome natural power obviously impressed the ancient Hawaiians who arrived from the Marquesas Islands and the Society Islands more than a thousand years ago. The Hawaiians shared many common elements of their cosmology with other Pacific peoples. Their legends and chants told of the creation and order of their universe and of the gods responsible for these supernatural activities. Unique, however, to Hawaii was the goddess responsible for volcanic activity. They called the goddess of the volcanoes Pele and paid her great respect. Although the Hawaiian's beliefs about Pele were mere attempts to explain natural phenomena, one legend accurately describes the geologic order in which the island chain was formed. Dr. Thomas L. Wright, senior scientist at the Hawaii Volcano Observatory, believes the Hawaiians accurately understood the relationships of the various natural phenomena associated with the volcanoes. According to Wright none of their concepts were really wrong. Centuries of astute observation and living close to nature enabled them to acquire a firm grasp of the dynamics of their island. They literally lived on volcanoes that were seldom dormant; Carbon 14 dating of lava flows indicates that most of the slopes of Kilauea and Mauna Loa and Hualalai have been inundated by lava flows since the arrival of the first Hawaiians.[2]

The ancient Hawaiians were very careful not to anger Pele and incur her wrath in the form of eruptions, earthquakes and tsunamis. Special sacrifices were thrown into Kilauea crater in careful ceremonies to appease her. Certain proof, however, that the new god of the Christians was more powerful that those of ancient Hawaiians came in December 1824 when the high chieftess Kapiolani, who had been converted to Christianity by the early Protestant missionaries, marched to Kilauea to challenge Pele. Ignoring the warnings of Pele's kahunas, or priests, she stood on the edge of the crater and dared Pele by throwing rocks into the crater instead of Pele's sacred red ohelo berries, sang Christian hymns, and prayed. When Pele did not respond, the Hawaiians knew that Jehovah was superior to Pele.

This was not the first time that Kilauea had interfered with Hawaiian history. In November 1790 a young King Kamehameha, who controlled the western side of the Island of Hawaii, was fighting King Keoua who controlled the eastern side. Keoua was attacking around the south end of the island and divided his army into three parties as they passed by Kilauea. Then one of Kilauea's two known phreatic, or explosive, eruptions occurred which totally wiping out the middle party of Keoua's army. The Hawaiians considered this proof that Pele had taken Kamehameha under her protection, and he went on to consolidate his rule over all the islands of Hawaii. What other national park has the good fortune to enjoy the protection of a goddess? Interestingly enough, Hawaii Volcanoes National Park Visitor's Center continues to inform the public about Madame Pele and warn them of her jealousy and vengeance. In a central showcase in the exhibit hall are views of a variety of sizes and types of lava taken away by mainland tourists and then returned because Pele had shown her wrath in the form of ill luck.

EARLY MAPS OF THE PARK AREA

Approaching Hawaii from any direction, European explorers could not help but be impressed by the massive mountains which they recognized as volcanoes. Mauna Loa and Mauna Kea are the main, and only, features in the interior of the island of Hawaii on the map produced by the expedition of Captain James Cook during his visits in 1778 and 1779. That map and another made under the direction of Captain George Vancouver in 1794 were the major sources for all other maps of Hawaii for decades. Archibald Menzies, a naturalist with Vancouver, made the first ascent and description of Mauna Loa in February 1794.

Not until 1823, however, was Kilauea visited and described by Europeans. In August of that year the Reverend William Ellis of the London Missionary Society, who had come to Hawaii after five years in Tahiti, toured around the entirety of the Big Island. Different versions of the map to illustrate his travels were made for the American and English editions of his book; they are the first maps to locate Kilauea. The 1825 American edition uses pictorial relief and overall is more primitive but illustrates the massiveness of Mauna Loa and Mauna Kea better. The 1827 English edition has more place names and includes 'burning chasms' but its use of hachuring presents a less accurate view of Hawaii's topography.

In 1825 Lieutenant Charles R. Malden, who accompanied Lord Byron in returning the bodies of Liholiho (King Kamehameha II) and Queen Kamamalu to Hawaii, visited Kilauea and drew the first detailed map. According to Gary Fitzpatrick in *The Early Mapping of Hawaii*, "Malden's plan is of such detail that later scientists have used it as the beginning point for analysis of the historic changes in the configuration of the crater." Malden's map of "the volcano Pele" is reprinted on the United States Geological Survey's 1986 map of Hawaii Volcanoes National Park.

The United States Exploring Expedition under the command of Charles Wilkes spent the winter of 1840-1841 in the Hawaiian Islands. The expedition's corps of scientists spent considerable time investigating the areas that would later be incorporated into the national parks. Wilkes himself led a party of sixteen Americans and two hundred Hawaiians from Hilo to the summit of Mauna Loa and set up camp on the east rim of the crater of Mokuaweoweo. The party spent almost three weeks in the blizzards and gales of the summit. Wilkes personally supervised the surveying of Mokuaweoweo using base lines, triangulation, and a theodolite. The expedition's map of Kilauea was made from observations taken following the return from Mauna Loa on January 16th and 17th. At Kilauea, a local missionary and guide for the expedition, Dr. Gerrit P. Judd, descended into the crater to collect some fine specimens of capillary glass called "Pele's Hair" from the inside walls of the crater. It began to erupt and Judd barely escaped, badly burned, with the aid of a Hawaiian friend. The expedition then explored the rift zone to the east of Kilauea all the way to the large lava flow of 1840 at Cape Kumukahi.

Wilkes and his officers tied their surveys throughout eastern Hawaii together on a general map at the scale of 1:191,300. It covers most of the area now in the Hawaii Volcanoes National Park and is a fine example of copper plate engraving. Relief is shown by hachures and other patterns depict vegetation, lava flows and other topographic features; it was printed in volume two of the rare *Atlas of Charts* to accompany the official report of the United States Exploring Expedition and is reprinted in Fitzpatrick and Moffat's *Early Mapping of Hawaii.*

While the maps of Hawaii by the United States Exploring Expedition are well executed, they do have flaws which are pointed out by Gary Fitzpatrick and Riley Moffat in *The Early Mapping of Hawaii.* Fitzpatrick notes that, whereas Wilkes' extensive measurements should have allowed him to construct a very accurate map of a large part of the Big Island, there are several major flaws. The misplacement of Kilauea and Hilo skew the distances and relationships of these landmarks.[3] Charles H. Hitchcock in his *Hawaii and Its Volcanoes* wonders why Wilkes did not allow the expedition's geologist, James Dwight Dana, to lead the exploration of the volcanoes. Dana was allowed to spend only one day at Kilauea then but returned 46 years later to correct all of Wilkes' errors.[4] Combined with the vivid paintings of the expedition's artist, Titian R. Peale, the maps provide a spectacular vision of the inferno.

Because it was not considered an active volcano, Haleakala on Maui did not receive the same close attention as did Mauna Loa and Kilauea. An expedition group, including Dr. Judd and Lorrin Andrews of Lahainaluna, ascended the mountain for a few days in March 1841 and a map was drawn by an expedition artist, Joseph Drayton. His distances and angles are estimated without benefit of base lines or triangulation, and while Drayton's map is the first of Haleakala it does not do it justice.

In 1846 the Reverend Chester Smith Lyman, a graduate of Yale's Sheffield Scientific School, visited Kilauea on his way to California, where he became a prominent surveyor. W. T. Brigham, in his memoir for the Bishop Museum, prints two versions of a map Lyman made showing the configuration of the crater in 1846. Thus, beginning with Malden's 1825 map, an excellent cartographic record of the volcanoes exists to illustrate their geologic dynamism. The floor of the crater has fluctuated up to one thousand feet as the pressures build and recede. This phenomenon is best documented by the maps that have been made of the crater over the past 164 years.

In 1864 a young Harvard graduate and scientist came to Hawaii to explore geology and botany before returning to teach at Harvard. William Tufts Brigham stayed in Hawaii until 1865 and in August of that year revisited Kilauea to make a proper survey and map using chains (fig. 13). He also established the magnetic declination for five points around the crater to show its radical variance over short distances. Charles Hitchcock, writing in 1909, claimed that Brigham's accurate map has been the basis of every plan published since that time. Brigham served as curator and then as director of the Bishop Museum from 1890 to 1918.

The lake of molten lava in the crater of Kilauea, called "The Fires of Pele," was becoming a famous tourist destination and the first Volcano House was built on the north rim of the crater. Since 1866 a visitor's book has been kept where visitors could record their impressions of the volcanoes. While many entries were the trite and mundane, observant and scholarly visitors included maps, sketches, and scientific data in their entries.

The first scientifically executed map of Haleakala was drawn by William DeWitt Alexander in 1869. At the time he was employed as President of Oahu College, now called Punahou School, but supplemented his income in the summer by surveying. He had learned the art of surveying from his father who taught it at Lahainaluna School on Maui and he later studied surveying at Yale. Alexander's map of Haleakala exhibited the highest quality of mapping done in Hawaii up to that time. In his survey he established a base line over a mile long on the crater floor, and to keep his triangulation accurate, he determined the magnetic declination of his triangulation points. The visual appearance of the map was enhanced by reproducing it lithographically rather than by copper plate engraving like Wilkes' map; lithography gave a more vivid picture than engraving. Popular Victorian

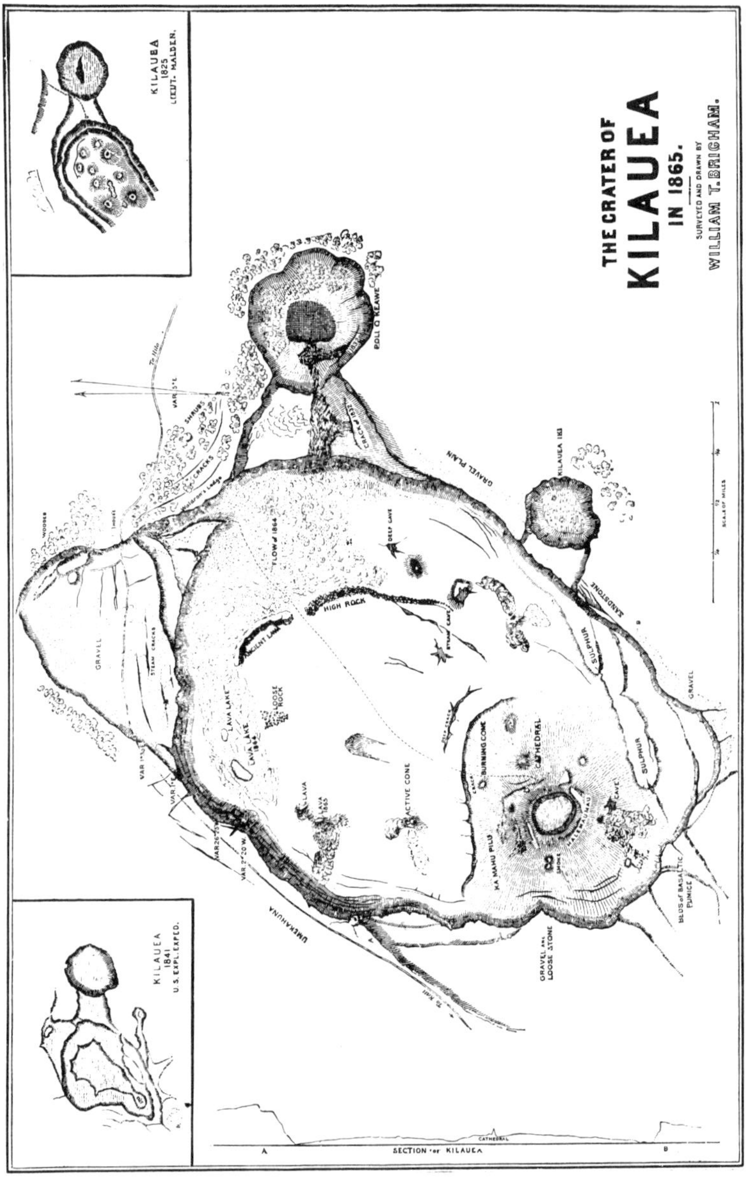

Fig. 13. Courtesy of Brigham Young University, Hawaii.

travel writer Isabella Bird complimented Alexander's map by saying that it gave "a far better idea of it (the crater) than any verbal description" and that the map illustrated Haleakala's "singular sublimity and magnitude upon one far more forcibly than a single visit to the actual crater."[5]

William Tufts Brigham's 1909 study of *The Volcanoes of Kilauea and Mauna Loa on the Island of Hawaii* reproduces a number of early manuscript maps of the volcanoes taken from the Volcano House visitor's books and other sources. Some also appear in Clarence Edward Dutton's 1884 study *Hawaiian Volcanoes* and Charles Henry Hitchcock's 1909 book *Hawaii and Its Volcanoes.* Three small plans that deserve mention are W. W. Hall's sketch of the Mokuaweoweo crater atop Mauna Loa in September 1873 and John Mortimer Lydgate's maps of Kilauea in January 1874 and Mokuaweoweo in June 1874. Lydgate also worked from time to time for the Hawaiian Government Survey and was hired to survey a road from Hilo to Kilauea. His entry in the Volcano House visitor's book dated June 24, 1874, reads: "The plan of the crater of Mokuaweoweo is from actual survey by triangulation. A base of 1876 feet was measured on the eastern side and from this, using a seven inch Queen transit, some twenty points were fixed indicative of the shape, topography, etc. of the crater."[6]

The Hawaiian Government Survey was established in 1870, and Professor W. D. Alexander, the man who had surveyed Haleakala so well, was appointed surveyor general. As part of the Hawaiian Kingdom's display at the Philadelphia Centennial exhibit in 1876, the Hawaiian Government Survey produced a monumental new map of the kingdom at the scale of 1:360,000. As insets on the map, the survey added plans of Kilauea, Mokuaweoweo, and Haleakala at the scale of 1:63,360 and graphically compared them with the exhibition's buildings in Philadelphia. The insets were based on the previously mentioned surveys of Haleakala by W. D. Alexander in 1869, Mokuaweoweo by J. M. Lydgate in 1874, and Kilauea by W. T. Brigham in 1865. They prominently displayed what Hawaii considered its most important attraction.

Professor Alexander's map of Haleakala was revised by S. E. Mann in 1884 (fig. 14) as the Hawaiian Government Survey compiled all of its surveys of lands on Maui into a large map of the entire island which was published in 1885. The Hawaiian Government Survey was also compiling its large scale surveys of the Big Island of Hawaii into a map of the island as a whole. This map, a preliminary edition at the scale of 1:240,000, was completed at the end of 1886. It was titled "Preliminary Edition" because the surveys in the districts of Puna and Ka'u, the very districts in which the future national park was to be established, were not completed. The 1886 map is important because it delineates and dates the recognizable historic lava flows on the island as well as the extent of forested lands.

As part of this effort on Hawaii, the Government Survey hired James McKinney Alexander, the Surveyor General's brother, and Joseph Swift Emerson to map the land boundaries on the frigid summit of Mauna Loa. Their story and map are recounted in Brigham's memoir and in James Dwight Dana's *American Journal of Science* of July 1888.[7] It must be remembered that Dana had been following Hawaii's volcanic activities since his visit with the United States Exploring Expedition in 1840-1841.

Kilauea erupted again in March of 1886 and the Government Survey sent senior surveyors and cartographers Joseph Swift Emerson and Frank S. Dodge to map the changes in the crater. Photographer J. P. Severin helped them lay out a base line of 3532 feet and they took measurements for their map, which appears in Brigham's *Memoir*, between March 24 and April 14. The final copy was drawn at the scale of 1:6,000.

The Hawaiian Government survey closely monitored changes in the volcanoes during the 1880s and 1890s. Maps to illustrate these changes, particularly those of Frank S. Dodge, are in the vault of the State Survey Division in Honolulu. Not only did Dodge execute detailed vertical maps but also horizontal sections to illustrate the changes in the crater which could fluctuate hundreds of feet in a day. Brigham's memoir shows that Dodge also drew his observations in the Volcano House visitor's book. Moffat and Fitzpatrick in their forthcoming *Early Hawaiian Maps* will reproduce and describe F. S.

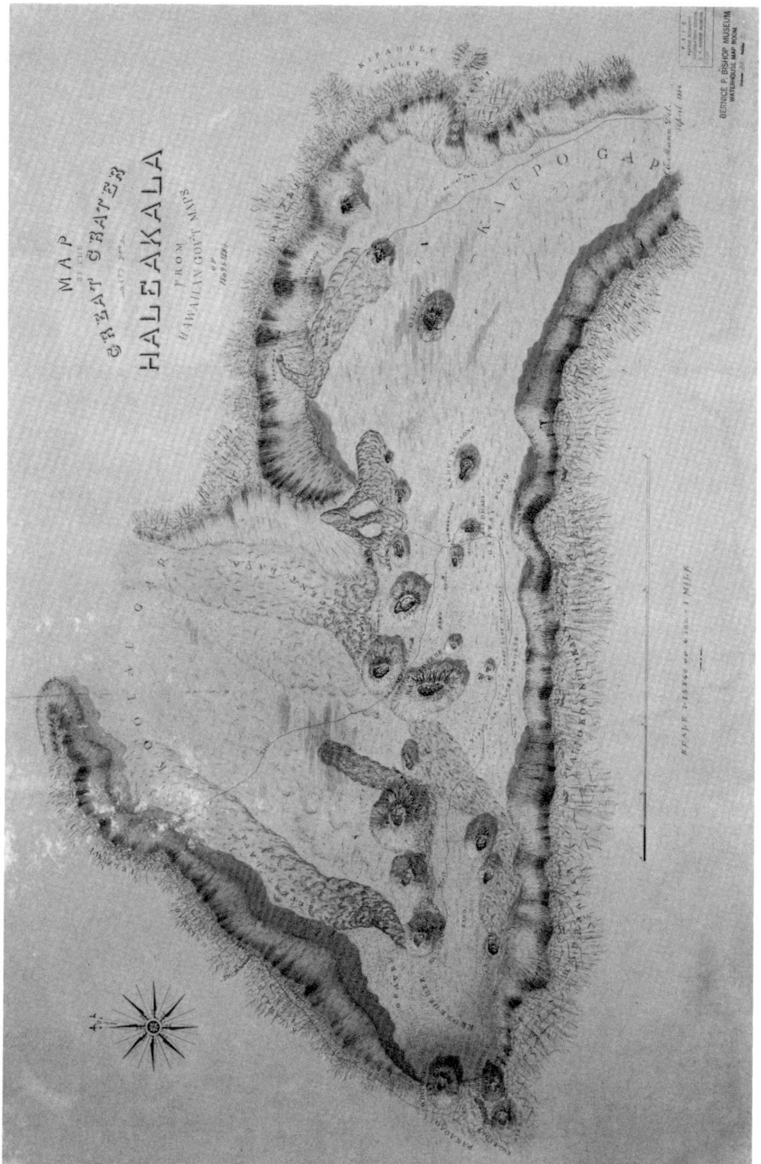

Fig. 14. Courtesy of Bishop Museum.

Dodge's 1886 map of Kilauea and his map and sections of Halemaumau showing change between 1886 and 1892. The observations of the staff of the Hawaiian Government Survey and copies of their maps continued to regularly appear in James Dwight Dana's *American Journal of Science.*

After 1890 the Hawaiian Government Survey stationed one of its experienced men in Hilo to coordinate work on that side of the island, especially the surveys of homestead land between Hilo and Kilauea. Thus being in close proximity to Kilauea and Mauna Loa, Erdman Dwight Baldwin closely monitored their activity and made a number of drawings and maps from 1892 to 1907. On December 26, 1906, he completed a plan that compared the fluctuations in Halemaumau, Kilauea's fire pit, in 1886, 1892, and 1906.

Charles H. Hitchcock in his 1909 book *Hawaii and Its Volcanoes* reports an interesting notion from Albert Brown Loebenstein, a government surveyor and land agent:

> Mr. A. B. Loebenstein of Hilo, is reported in the Washington Post of Feb. 6, 1906 as saying that Halemaumau has moved south 1,783 feet during the past thirty years. In 1873 when surveys were made in connection with the leasing of the land, this crater stood within the area of Keauhou, the boundary line passing through its center. Thirty years later a new survey was made which proved that the floor of the crater had crossed the boundary line into Kapapala. "It should be understood", he says, "that the inner crater of Halemaumau had not disappeared in one place and broken out in a new spot, but that it has worked itself along the floor of the outer crater foot by foot for a third of a mile".
>
> These remarks were made in connection with a suggestion that the Government should take possession of the volcano and its surroundings for a National Park. Incidentally it may be a question of ownership, and concerning that we have the reported remarks of Mr. F. S. Dodge, Superintendent of the Bishop Estate.
>
> The shifting of the lake of fire does not invalidate any claim we may have on it. The line of the Bishop estate is tied to the center of the lake, and our boundaries move along with it. The description of the boundary shows that the line runs from well known points

> on the bluff 'to the center of the lake of fire' and it is a well recognized fact in the Territorial Courts that a recognizable fixed point has precedence over distances and bearings. Thus, no matter where the lake of fire may wander to, it drags our line along with it.[8]

In 1909 John Mortimer Lydgate, an early government surveyor who had become a minister and civic leader on Kauai, returned to Kilauea from July 5th to 14th and drew a map of Halemaumau which appears in Brigham's memoir. Though there is evidence of a base line Lydgate disclaims great accuracy but states that his plan is substantially correct.

The United States annexed Hawaii in 1898. Hawaii already had been well mapped by the Hawaiian Government Survey under the direction of Professor W. D. Alexander, and the mapping activities of the Hawaiian Government Survey were split between the Hawaii Territorial Survey and federal agencies such as the United States Coast and Geodetic Survey, the United States Geological Survey, and the United States Army Engineers.

In 1901 the newly organized Territorial Survey revised the map of the Island of Hawaii that the Hawaiian Government Survey had published in 1887. This new edition was enlarged to 1:190,080 and included the completed surveys in the districts of Puna and Ka'u. This map appeared in the Serial Set in 1901 and again in 1906 with an overprint of major economic land usage.

A figure remembered for his early and significant work with volcanoes is Dr. Thomas A. Jagger. Believing that the only way to know a volcano is to live with it, "he built his home on stilts, wedged into a crack in an immensity of dark volcanic rock on the southeast flank of the island of Hawai'i."[9] Supported initially by the Massachusetts Institute of Technology, where he taught, Dr. Jaggar visited Hawaii in 1909 seeking to identify the ultimate location for a volcano research station. Honolulu businessman Lorrin A. Thurston assisted Jagger in organizing the Hawaiian Volcano Research Association and M.I.T. assigned Jagger to Hawaii on January 1, 1912. The association assisted in the support of Jagger and the Hawaiian Volcano Observatory which today is part of the Geologic Survey. The observatory monitors volcanic activity in the park and continues to support research activities. Adjacent to the Observatory on Crater Rim Drive is the Jaggar Museum operated by the National Park Service.

One of the first projects initiated by Jaggar was the mapping of the volcano area around Kilauea. The map would serve as a basis for the measurement of changes in the fire pit. With the backing of Hawaii's Governor Frear, Claude Birdseye, chief topographical Engineer of the Geological Survey after World War I, and Albert Burkland, topographic Engineer in charge of Hawaiian operations of the Geological Survey, were engaged to execute the map. The triangulation and plane-table survey was conducted in 1912 and the survey published the special map titled *Proposed Kilauea Volcano National Park* (fig. 15). This map was printed as a special "advance" sheet in one color. The area included Kilauea, Kilauea Iki, and the Chain of Craters rift zone east to Napaau as well as a portion of the Ka'u Desert. One feature typical of the days of ground-based mapping is an index map at the bottom of the sheet which shows the areas the two surveyors, Burkland and Birdseye, were individually responsible for. Printed at a scale of 1:31,680, or two inches to the mile, the map indicates crater contours and identifies historic lava flows, lava trees, "kipuka" (areas surrounded by flows), and caves.

From the mid 1800s, tourists came to Kilauea despite the rigors of travel. Early tourist maps, however, do not abound. One of the earliest, *Island of Hawaii, Territory of Hawaii, Map and Guide*, dated 1910 affirms the site's popularity:

> By reason of ease of access, as well as an account of superlative importance, Kilauea is the destination of a large part of the tourists who visit the island. A week or more may be profitably spent in the vicinity of the great pit . . . A horseback ride of some six or seven miles takes one to the very interesting Seven Craters, one of which is 1500 feet deep. Within the past year a splendid automobile road has been completed right into the crater of Kilauea, by which it will be possible to ride to the very brink of the inner fire-pit.
>
> Kilauea may be visited easily even by a person in feeble physical condition; and the fact that there has never been a single fatality in connection with the crater's activities, is reassuring the timid.[10]

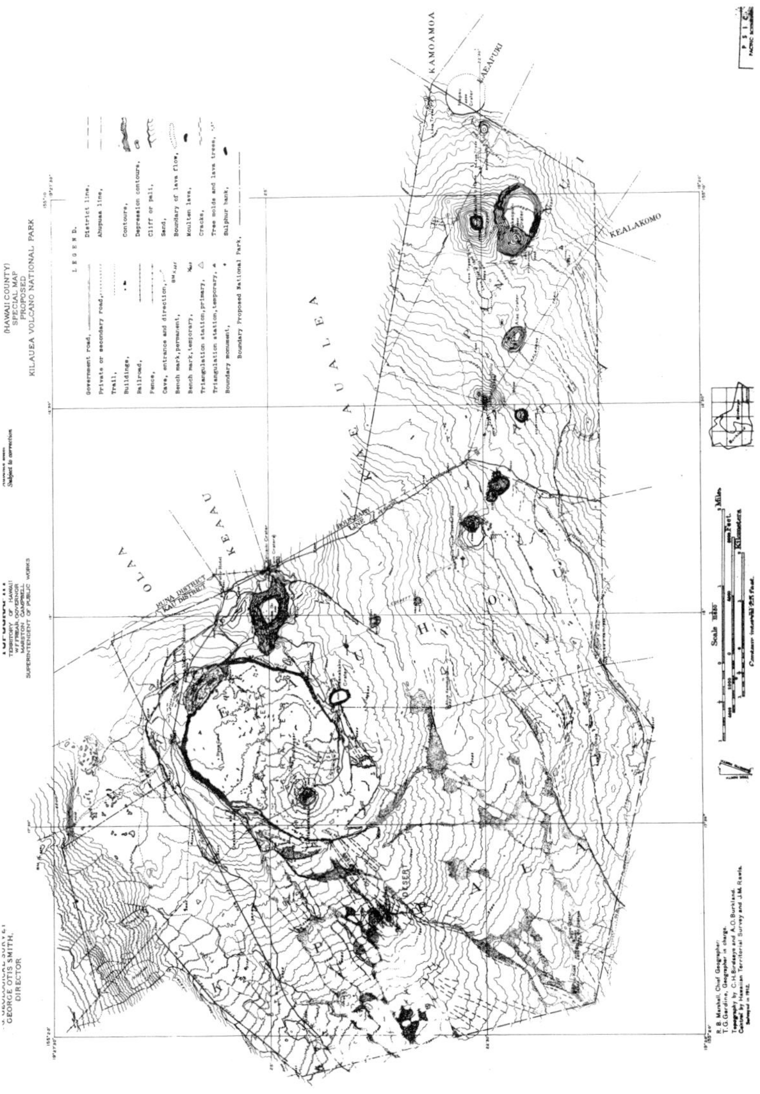

Fig. 15. Courtesy of Bishop Museum.

The guide includes two maps by H. E. Newton. The larger map of the entire island marks the main automobile roads up to Kilauea. Along the route from Waiohinu to the crater should be noted a remnant of the early days, a halfway house. This and others like it were constructed for the convenience of the tourist on the two day trip up to the volcano. The description of the house given by Isabella Bird, however, in her book *Six Months in the Sandwich Islands* might have dissuaded some tourists from attempting the trip ". . . as I was hardly able to sit on my horse owing to fatigue, I consoled myself by visions of a comfortable sofa and a cup of tea. It was with real dismay that I found the reality to consist of a grass hut, much out of repair, and which, bad as it was, was locked."[11]

In 1913 John M. Akau made a map of the Island of Hawaii at the scale of 1:365,000 based on the 1901 Hawaii Territorial Survey map. On the verso was a map of Kilauea and vicinity at the scale of 1:94,000 prepared from the 1912 surveys of the United States Geological Survey. The map was copyrighted by Henry Wadsworth Kinney and was published in his *Island of Hawaii* which he produced for the Hilo Board of Trade. It is an excellent early example of a Hawaiian tourist promotional map.

TERRITORY OF HAWAII MAPS

One of the first maps of the Hawaiian Islands to show the new Hawaii National Park was *The Hawaiian Islands* published by the Hawaii Territory Survey in June 1919. Printed in black and white at the scale of 1:720,000 on thin paper with a wax finish, the map was drawn by H. E. Newton in a style typical of the times. Relief is represented by hachures. Hawaii National Park is shown as three small, unconnected parcels around Kilauea, Mauna Loa, and Haleakala. In addition to the eight major islands, there is an inset with an enlargement of the three main craters reminiscent of the Hawaiian Government Survey's 1876 Centennial Map of Hawaii. Place names are abundant as well as descriptions of the terrain: sandy plain, gravel and large boulders.

The Hawaii Territory Survey produced a separate map of the Island of Hawaii at the scale of 1:240,000 in December of 1928. The 1928 map shows the Hawaii National Park which by that time included the portion connecting Mauna Loa and Kilauea. The coastline of the park extended from Na Puu o na Elemakule to Keauhou.

Other maps published by the Territory of Hawaii include a large *Topographic Map of the Island of Hawaii* at a scale of 1:125,000 by the Division of Hydrography published in 1944. This map was based on surveys done by the Geological Survey, but compiled by the National Park Service, which is surprising since this is a map of the entire island. Revisions are by the noted Hawaiian geologist Harold T. Stearns, who, like Jaggar, had a fascination with volcanoes. A witness to the 1924 Kilauea eruption in which no magma was involved yet blasted out 10 ton rocks, Stearns said, "Watching that mountain blow its top was one of the most exciting moments of my life. I knew right then that I was going to have a lot of fun in the islands."[12] He is responsible for much of the original geologic mapping of the Hawaiian islands. For about thirty years, from the 1920s through the 1940s, Stearns rode thousands of miles on mule or horse into every major valley and scaled each mountain range in the Hawaiian chain to collect data. In addition to his geologic maps, he authored the classic texts *Geology of the State of Hawaii* and *Guide to Points of Geologic Interest in the Hawaiian Islands.*

In 1945 Stearns and another renowned geologist and volcanologist, Gordon A. MacDonald, worked for the Territory of Hawaii's Division of Hydrography to produce the *Geologic and Topographic Map of the Island of Hawaii.* Based on the Stearns map just mentioned, this particular map indicates locations of wells, springs, tunnels, and shafts. A legend explains the color coding indicating geologic age and composition of the volcanic series. At the bottom of the map are cross sections showing elevations of the volcanoes and extent of the intrusive complex where lava accumulates underground. This map was the most detailed map of its kind up to this time.

UNITED STATES GEOLOGICAL SURVEY

After Hawaii became a territory of the United States in 1900, the United States Geological Survey assumed a key role in the mapping of the islands. The first series maps of the island of Hawaii was produced because there was an acute need for topographical maps to understand surface and ground water resources. These quadrangles, published from 1912 to 1930 at a scale of 1:31:360 on seventy-three sheets, were labeled "advance sheets" and were printed in brown ink. Each covers 7.5 minutes of longitude and latitude, and they indicate contours, date lava flows, and mark several curious sites including a prison camp on the Kilauea quadrangle. This consisted of four structures on what is now Namakani Paio campground. The camp housed prisoners working on the park's road project. Another feature to note is a stone corral inside the crater, which was probably used to keep the horses which ferried visitors to and from the volcano's fire pit. Both of these features are indicated on the 1:62,500-scale, 15-minute maps which were constructed from the 7.5-minute advance sheets. These maps also indicate by straight lines the cracks and fissures common in the park area.

In 1927 the Geological Survey revised and enlarged its map of the "proposed" Hawaii National Park, now a reality, and printed it at the scale of 1:62,500. The boundaries included Moana Loa, the piece connecting it to Kilauea, and the slope continuing on down to the coast between Na Puu o na Elemakule to near Keauhou. The map also marks the location of wells or tanks, aqueducts and pipes, telephone lines, water and shelter areas, and park headquarters near the Volcano House. Kilauea Military Camp located near the park headquarters was begun as a training camp for the National Guard in 1916 and is today a military rest and recreation facility, probably the only one of its kind in a national park. Boundary changes from 1916 to 1928 are encompassed in the 1933 edition of the map.

The Geological Survey remapped the Hawaiian Islands from 1952 to 1967 at the scale of 1:24,000 on sheets generally covering 7.5 minutes of longitude and latitude. These maps show remarkable detail and are generally the most detailed maps available for many areas. Portions of the park appear on nine of sheets, bearing the name Hawaii Volcanoes

National Park which was given in 1961 when the Haleakala section on Maui gained separate status as a national park. The sheets covering the Kilauea area were mapped in 1963.

The number of sheets needed to cover Hawaii Volcanoes National Park reflects the vast size of the park, 891 square km. or 344 square miles. On close examination, one notices the variety of topographical features and habitats. Besides craters and cracks associated with volcanoes, there are cliffs called "pali," a result of shifting and dropping of land due to earthquakes. The various colors, from dark green on the northeast side of Kilauea Crater to brown on the southwest side indicate the rapidly changing land cover, from lush rain forest, woodland, to scrub and desert. In contrast to the rain forest around the park headquarters, the land just southwest of Kilauea Crater is a stony desert and is said to be so much like the moon that part of the Apollo astronauts' training took place there.

Between the production of this initial series of 7.5-minute topographical maps and the 1978 orthophotoquads from which the complete revision was produced from 1980 to 1984, much volcanic activity took place leaving its mark in the form of lava flows, enlarged craters, and deforestation. In addition, as Hawaii's molten lavas flow to the sea, hundreds of additional acres are added to the island. The earthquakes that often attend an eruption, however, can be subtract acreage when coastlines subside. For example, the January 1960 eruption at Kapoho added 500 acres to the island at Cape Kumukahi. Later, on November 29, 1975, a major earthquake accompanied an eruption. Parts of the coast south of Kilauea sank into the sea three to twelve feet and a fifty foot high tsunami washed inland. The change in the coastline can be seen by comparing the coastline from Keauhou Point (on the Makaopuhi Crater quadrangle) to Kalue (on the adjoining Ka'u Desert quadrangle) on the 1963 edition with the 1981 edition. Hawaii Volcanoes National Park is probably the only park capable of expansion or reduction without an act of Congress!

Military cartographers have had a great interest in Hawaii and were active in the islands before the Geological Survey began operations in 1910. In 1942 the United States Army Map Service (AMS) produced a series of 1:250,000 intermediate scale maps for Hawaii. This series was revised in 1951 by the AMS and published as its series W532. A separate edition of this series showed relief by hill shading as well as by contours. A limited revision of some sheets was made in 1962 and a full revision of the two Big Island sheets, combining them into one sheet, was made in 1975 by the Geological Survey which had assumed responsibility for this series nationwide. More recently, in 1988 the Defense Mapping Agency, successor to the Army Map Service, reformatted the 1:24,000-scale 7.5-minute topographic quadrangles of the Big Island into 1:50,000-scale 15-minute sheets with metric contours. Comparing all these various editions shows the numerous rerouting of roads in the park to accommodate the lava flows that occasionally cover them.

A 1:100,000-scale county based series of maps was initiated by the Geological Survey in 1975 and planimetric coverage of the state of Hawaii was prepared in 1980 but not printed. However, a topographic edition of Hawaii County in three sheets was published. The planimetric version served as a base for a series of maps showing land usage in Hawaii during the period 1976-1978. Besides indicating economic usage by man they also show the natural state of the land cover for that time period.

The most recent single sheet map of Hawaii Volcanoes National Park appeared in 1986. The 1:100,000-scale map by the Geological Survey shows the entire park including current lava flows begun by the Kilauea eruption in 1983. On the reverse side is a 1:24,000-scale shaded relief map of just the Kilauea crater area. Two editions were printed: the regular edition has shaded relief with northeast illumination and a special edition with northwest illumination. Description of the various smaller craters, the geology of Hawaii and a glossary of Hawaiian place names in the park is printed on the verso of the map.

Besides the topographic maps produced at various scales, the Geological Survey has also created maps to show the geology and ground-water resources of the national park area. The oldest published study is H. T. Stearns' and W. O. Clark's *Geology and Water Resources of the Ka'u District, Hawaii* which came out in 1930 as Water Supply Paper 616 with a map at the scale of 1:62,500. *Bathymetric, Topographic, and Structural Map of the South-Central Flank of Kilauea Volcano* by J. G. Moore and D. L. Peck was published in 1965 as

Miscellaneous Investigations Map I-456. Then appeared a series of four geologic maps of Hawaii Volcanoes National Park based on the existing 7.5-minute topographic quadrangles: D. W. Peterson's *Geologic Map of Kilauea Crater Quadrangle, Hawaii* published in 1967 as Geologic Quadrangle Map GQ-667; G. W. Walker's *Geologic Map of the Kau Desert Quadrangle, Hawaii* produced in 1969 as Geologic Quadrangle Map GQ-827; G. A. MacDonald's *Geologic Map of the Mauna Loa Quadrangle, Hawaii* produced in 1971 as Geologic Quadrangle Map GQ-897; and G. A. MacDonald's *Geologic Map of the Kipuka Pakekake Quadrangle, Hawaii* published in 1971 as Geologic Quadrangle Map GQ-957.

To celebrate the diamond jubilee of the Hawaiian Volcano Observatory in 1987 the Geological Survey gathered a number of papers reflecting the breadth of research on Hawaii's volcanoes into its 1667-page Professional Paper 1350, *Volcanism in Hawaii.* These papers include numerous maps detailing all aspects of earth science research on Hawaii's active volcanoes. Of special interest are maps of the undersea volcano Loihi and maps colored to show the date and extent of Kilauea's lava flows over the past few thousand years. They show that almost all the land in the Puna district has been inundated by Kilauea lavas within the last 500 years.

The Geological Survey has produced a number of maps depicting geologic and seismic attributes of the region. Miscellaneous Investigations Map I-1611, *Earthquake Map of South Hawaii, 1968-1981,* charts the large amount of seismic activity in the southeast quarter of the Island of Hawaii and the offshore waters over a very generalized base map. The epicenter and depth of the earthquake are color-coded for the period 1968-1981.

MF-1845-B, *Apparent-Resistivity Map of the Rift Systems of Kilauea and Mauna Loa Volcanos, Island of Hawaii, Hawaii,* displays the magnetism of rocks associated with the rift systems of the Kilauea and Mauna Loa volcanoes, and MF-1845-A, *Aeromagnetic Map of the Rift Systems of Kilauea and Mauna Loa Volcanos, Island of Hawaii, Hawaii,* looks at low level aeromagnetic data over the same area. Both data sets are overprinted on the 1:100,000 topographic base.

MF-1368, *Structural Map of the Summit Area of Kilauea Volcano, Hawaii*, by Patrice de Saint Ours in 1982, examines the surfacial structural geology of the Kilauea summit area. Added to the 1:24,000 topographic base are data on geologic hazards such as faults, eruptive fissures, and cracks. This information should be of interest to visitors and hikers as well as scientists.

I-1908, *Generalized Ages of Surface Lava Flows of Mauna Loa Volcano*, by J. P. Lockwood and published in 1988, portrays one of the Geologic Survey's most recent studies. Similar to Robin T. Holcomb's detailed study of the eruptive history of Kilauea, published with many small maps as Chapter 12 of Professional Paper 1350, I-1908 shows, by color-coding, the age of lavas emanating from Mauna Loa based on Carbon-14 dating back more than 4,000 years. This data is printed on a generalized 1:250,000 topographic base.

This and other data is helping the survey to compile a new geologic map of the entire island. This map is being compiled now and will be printed on the survey's three sheet topographic base map of the island of Hawaii at the scale of 1:100,000. The volcanic activity that the national parks in Hawaii were created to protect and study have been the focus of two maps of Hawaii detailing the seismic activity of the state and the geothermal potential. *Geothermal Resources of Hawaii* was published in 1983 using the Geologic Survey's 1:500,000-scale topographic map as a base with an inset of the Kilauea area at 1:125,000. The data was compiled by the Hawaii Institute of Geophysics at the University of Hawaii and produced by the National Geophysical Data Center in Boulder, Colorado. Susan K. Goter's *Seismicity of Hawaii, 1962-1985* appeared in 1988 as the Geological Survey's Open File Report 88-285. This map is something of an anomaly in that it is stunningly printed in vivid colors on a base map that Raven Maps & Images developed from Geological Survey material rather than the black and white, often preliminary, style usually seen in Open-File Reports.

An annotated bibliography of all geologic and seismic studies of the volcanoes of Hawaii from 1779 to 1955 including maps was published in 1989 by University of Hawaii Press. Compiled by Thomas L. Wright, senior scientist at the Hawaii Volcano Observatory, and Taeko Jane Takahashi, librarian at the observatory, and titled *Obser-*

vations and interpretation of Hawaii Volcanism and Seismicity 1779-1955: An Annotated Bibliography and Subject Index, it includes illustrations of several important historical maps. A follow-up volume by Tom Wright is *Hawaii Volcano Watch: A Pictorial History, 1779-1991* published by University of Hawaii Press in 1992.

PARK SERVICE AND GENERAL LAND OFFICE MAPS

Like other national parks and monuments the national parks in Hawaii created general maps and brochures to issue to visitors. In the 1920s the Park Service maps were clear black and white line maps with no relief. By the 1950s the maps used various inking patterns to indicate different areas and ages of lavas and appear to be based on the maps of Stearns and MacDonald. In the late 1960s the Park Service tended toward abstract color designs and cartoonlike illustrations and maps. While colorful, they were not as precise as the earlier maps. In the early 1970s much of the graphics and artwork was still somewhat abstract but the maps were clearer; these maps had better line work and used shaded relief. The need for clear, precise, yet visually appealing and colorful maps appear to be well served by those currently produced. The present National Park Service map for Hawaii Volcanoes National Park shows gross relief with contour lines and uses various solid colors to show and date recent lava flows in the Kilauea area. The map for the Haleakala National Park brochure uses the Geological Survey's 1:62,500 contour map of the park area as a base on which to place improvements and trails. The park boundaries are outlined in the latest edition of the National Park Service's Hawaii Volcanoes National Park map. A more detailed map of land acquisition for the period 1916 to 1979 is included in the National Park Service draft land protection plan of 1985.[13]

From 1900 through 1929, the General Land Office (GLO) of the Department of the Interior annually produced a map titled *Map of the Territory of Hawaii* at the scale of 1:760,320 which was tipped into the Governor's Report to the Secretary of the Interior each year. Along with the eight major islands, there is an inset of the Hawaiian archipelago in the lower left corner. This map from 1918 on also features a table giving the distance from Honolulu to various other Pacific ports.

The growth of Hawaii National Park boundaries during this time can be found on these GLO maps. From 1901 to 1918, no national park is shown, although the Volcano House, half-way house and lava flows arc marked. In 1919, the map outlines and shades in pink three distinct parcels: Haleakala on Maui, and Mauna Loa and Kilauea on Hawaii. The next change in park boundaries appears on the 1923 edition. It shows the Kilauea portion extending down to the coast. On this map, however, the half-way house has disappeared and the carefully drawn lava flows formerly indicated by a stone pattern no longer appear. The last GLO map in 1929 shows the Mauna Loa and Kilauea sections as one parcel.

TOURIST MAPS

In the 1920s, the Hilo Automobile Club published two maps on one sheet. The larger map of the Island of Hawaii indicates tourist attractions, hotels, roads, trails, and even railroad tracks. The cartographer, W. H. Barringer, does not indicate Hawaii National Park on the map. The smaller map on the bottom by T. Boles is dated 1923 and is titled *Airplane View of Hawaii National Park Showing Volcanoes, Craters, and Recent Lava Flows on the Island of Hawaii.* It is a low oblique view from the south; before the advent of airplanes these views were commonly known as "birds' eye views." This map places the Hawaii National Park right in the forefront while the rest of the island and the other islands north to Oahu recede in the background.

In 1949 the Hawaii Natural History Association published *Self Guiding Map of Hawaii National Park.* Besides detailed description of places and features in and out of the park, it contained four maps, including one of the Kilauea caldera and another of the entire park area. The maps were compiled by James Y. Nitta of the Geological Survey who effectively utilized a stippling technique to indicate relief. Information on the map such as trails, roads, and tourist sites remains a record of the history of the continuously changing environment. For example, the map includes the Devil's Throat which was formed in 1921. It is a pit crater and was described in 1923 by W. T. Sinclair as a pit 300 feet deep, 222 feet in diameter at the bottom, and 35 feet in diameter at the top.[14] In addition, the guide lists the dates of eruptions of Mauna Loa and Kilauea up to 1949 and contains a guide to meanings of Hawaiian names and geologic definitions.

Today, tourist maps of the Hawaii Volcanoes National Park area, often mediocre in quality and appearance, abound. They are standard fare in the many guide books, brochures, and throwaway literature. Notable exceptions are the popular maps of Hawaii published by the National Geographic Society in 1960, 1976, and 1983.

James A. Bier, cartographer for the *Atlas of Hawaii* published in 1973 and revised in 1983 by the University of Hawaii Press, has produced excellent sheet maps of five of the Hawaiian islands. His map of the Big Island of Hawaii first appeared in 1976 with revised editions in 1979, 1983, and 1988. Bier's distinctive and attractive cartographic style and his use of the approved spellings and diacritics in Hawaiian place names contribute to the popularity of the maps among both tourists and residents.

In 1988 a new entry in the Hawaiian map scene appeared. Nelles Verlag of Munich, West Germany, published maps of the individual islands of Hawaii in its world-wide program of mapping exotic tourist destinations. They are pleasingly and recognizably done in the German style in terms of colors, linework and design. Without Bier's resources at the University of Hawaii, the accuracy has suffered slightly but it should improve with future editions. Nelles' Big Island map is printed at the scale of 1:330,000 with an inset of the Kilauea park area at 1:125,000.

HALEAKALA NATIONAL PARK

In time it became apparent that the administration of a park located on two separate islands was somewhat difficult to administer. In 1960 Roger Ernst, then Secretary of the Interior, stated in a letter to committees of both the House and the Senate:

> Haleakala's unusual geological history and inspirational scenic qualities makes it worthy of recognition as a separate unit of the

> National Park System. Evidence of its popularity is the fact that there were 56,900 visits to the area in 1958 . . . The other sections of the Hawaii National Park, a larger area and the site of park headquarters, is a distance of about 150 miles by land and air and water travel. There is no more reason for these two widely separated units to be designated as one national park than there would be, for instance, to have Yosemite and Sequoia National Parks combined.[15]

Haleakala National Park was thus designated as a separate park effective September 13, 1961. The former Big Island portion of the park was renamed Hawaii Volcanoes National Park because the foremost interesting feature was the ever present active volcanoes. The name change was approved September 22, 1961.

Territory, State, and Federal Mapping

Maps of Haleakala National Park, unlike those of Hawaii Volcanoes National Park, do not show a great deal of change in topography from one edition to another. Indeed, volcanologists describe Haleakala as dormant; it has not erupted in 200 years, though it is fully capable and even expected to erupt again. For the moment, however, peace reigns, and Haleakala National Park attracts visitors for its quiet beauty and habitats of plant and animal life found nowhere else in the world.

Haleakala is the world's largest volcanic caldera and one of the state's major wilderness areas. Until 1935, when the first road was constructed to the summit, the number of people able to visit the park was fairly well limited to the hardy and spartan wilderness seekers and physically fit scientists. The would-be tourist in 1929 looking for access to the summit of Haleakala would find on the Hawaii Territorial Survey's 1:60,000-scale *Map of Maui*, published that year, only a horse trail ending at Haleakala Rest House at the 9325 foot level.

The Geological Survey's 1930 edition of its 1:62,500-scale *Topographic Map of Maui*, (from surveys done from 1912 to 1925), while beautifully illustrating the size of the crater, also shows that access is only by trail. Later reprints in 1942 and 1948 fail to show any road even though the National Park Service completed the Haleakala Highway to the summit in 1935. Compared with the 1919 territorial map, the 1930 edition shows the reduction of the park from 21,150 acres to 17,130 because of a congressional decision to draw the park's boundaries along the crater rim.

The first Geological Survey map series of Maui came out as eighteen advance sheets at a scale of 1:31,680 from surveys made in 1922 to 1925. Produced simply, with contour lines in brown and place names in black ink, the maps give a fine representation of Haleakala. Noting the extensive gulches along the north and east slopes, one can easily see why the Hawaiians developed the trail system within the caldera to get from one side of the island to the other. Well established and in use before the caldera became a part of the Hawaii National Park, the main trails, Sliding Sands, Halemauu and Kaupo, amount to about thirty-six miles and appear on maps as early as 1869. Based on the above surveys, the Geological Survey published a series of five maps for Maui at a scale of 1:62,500 in 1925. Most of the park, which included at this time only the caldera area, appears on the Haleakala quadrangle.

In 1939 the Territorial Highway Department published its *Base Map Island of Maui, Maui County*. This map clearly delineates the road completed four years earlier through the park to Red Hill. Although there are few place names, the map shows the well established trail system and some topographic features. The Army Map Service's 1:250,000 scale map that covers Maui as part of its Series W532, compiled in 1951, also shows the new road to the summit.

One of the first United States Geological Survey maps to give evidence of such a major improvement to the park was the revision of its 1:62,500 scale map which was published in 1957. The new road increased visitation to the park from a few hundred to over 16,000 the year after it was completed. By this time however, the Rest House shown on earlier maps had been torn down. Instead, cabins named Holua, Paliku and Kapalaoa, all in the crater, are indicated. These were built by the Civilian Conservation Corps in the late 1930s and are still used by hikers today. Also on the map is Hosmer Grove Campground, named for Ralph Hosmer, first territorial forester

(1904-1914). He planted a stand of exotic conifers from North America, Japan, Australia, and the Himalayas at this site. Located at the 6500 to 9000 foot level, these exotics thrive and are one of the many unique features in Haleakala National Park. This popular map was quickly revised in 1961 to show the new name of Haleakala National Park. Later printings of this same map also show the addition of the 9,000 acre Kipahulu extension. The AMS 1:250,000-scale map of Maui which was taken over by the Geological Survey was also revised in 1962.

The first published 7.5-minute topographic quadrangles of Haleakala appeared in 1957 and bear the old name, Hawaii National Park; they were completely revised in 1983. Missing on this map at the entrance to the park just off the highway is the Silversword Inn, shown on the 1957 Kilohana quadrangle. Unfortunately, like the park concessions that preceded it, it also failed and closed in 1961. Silversword is the name of a trail within the crater on the Kilohana quadrangle as well and the name of the Hawaiian plant "ahinahina" which grows only on Haleakala. A relative of the sunflower, it is uniquely adapted to the crater environment with its leaves covered with fine hair to reflect the sunlight and retain moisture. Once in its lifetime after about twelve years of growth, it blooms.

After the Haleakala Highway was complete, Red Hill at the southwestern end of the park became easily accessible, and the undeveloped Federal land made this area attractive to the military during World War II. As a result, Red Hill was the site of several military attempts to establish communication stations and "unspecified defense installations." Remnants of this military use of the park appear on the summit of Red Hill on the 1957 map. Today, however, no non-park installation exists within the park boundaries. As seen on both the 1957 and 1983 editions of the Kilohana quadrangle, Science City occupies an adjacent area just outside the park around Kolekole Peak, but accessible only through the park.

Two important changes occurred between 1957 and 1983 when these series were published. One is the passing of the National Park bill in 1961 which designated Haleakala as a separate park and led to the Haleakala National Park name appearing on the 1983 maps.

The second change reflected on the 1983 maps is the extension of the park's boundaries to include Kipahulu Valley down to the coastline, a distance of about 8 miles from Haleakala's 7,000 foot level to the sea. These enlarged boundaries shown by the Geological Survey include additions that have not been firmly settled by the National Park Service, the Nature Conservancy, and other private landowners. Meaning "fetch from exhausted gardens," Kipahulu is one of three isolated ancient Hawaiian districts on this side of Maui. Residents of these three districts, Kipahulu, Kaupo and Kahikinui, enjoyed the benefits of owing little allegiance to outside chiefs. Even today, this area remains in near pristine condition because of its isolation. Access is by sea or by a narrow, winding road. Recently a proposal to build a tramway up the west slope on private land to the summit on park land was defeated by local voters.

Haleakala Park boundaries are drawn on current quadrangles to include the proposed acquisitions even though the park has not yet been given title to the land. More accurate boundary maps can be found in the 1986 *Land Protection Plan for Haleakala National Park* done by the National Park Service. Upon acquisition of the seven remaining parcels, the National Park Service can then issue an up-to-date official map of the Haleakala National Park.

Harold Stearns, working for the Hawaii Division of Hydrography compiled a *Geologic and Topographic Map, Island of Maui, Hawaii* in 1942. Similar to the map the division published for the Big Island, this map shows wells, springs, tunnels, shafts and an explanation describes the color code for sedimentary rocks. There is a description of East and West Maui rocks of the various volcanic series. Below the map is a cross section of the island showing elevation and geologic layering. Stearns used the 1927 Geological Survey topographic map for his base and added the road to the summit of Haleakala as well as the colored geologic overlay.

Long-time Hawaii volcanologist Gordan A. MacDonald mapped the geology of the Haleakala crater area for the Geological Survey and published the study as Miscellaneous Investigation Map I-1088, *Geologic Map of the Crater Section of the Haleakala National Park, Maui, Hawaii*, in 1978 at the scale of 1:24,000. A more unique and interesting map is Dwight R. Crandell's *Potential Hazards from Future Volcanic Eruptions on the Island of Maui, Hawaii* published by the Geological Survey as I-1442 in 1983. The map is a reduction of the 1:62,500 topographic sheet to 1:100,000. On this base Crandell has identified areas that have been inundated by lava flows over the past 200 to 2,000 years. Even though Haleakala's last eruption occurred in 1790 the intervening years are but a moment, in geologic time, and the volcano is still very much classified as active with the probable potential for more eruptions. Haleakala's eruptive history and the potential hazards from future volcanic eruptions, especially near the extensive Makena, Kihei, and Hana resort developments, should be of interest to planners.

A land use and land cover map based on the Geological Survey's 1:100,000-scale county maps shows economic land usage and the natural land cover of Maui during 1976-1978 by use of a numeric code. Most of the Haleakala National Park areas in the crater consist of bare exposed rock, with some range and pasture land on the north side. In contrast, most of the Kipahulu area is shown as forest cover.

Tourist Maps

Acquisition of the Kipahulu extension can be seen through the tourist maps by James A. Bier. The first edition published in 1976 shows the park's Kipahulu extension donated by Laurence Rockefeller as a narrow corridor down to the coast including the Ohe'o Gulch area of the Seven Sacred Pools. Later editions published in 1979, 1983, and 1988 show the park as it is today. The extension, now labelled "Scientific Research Reserve," was acquired through major efforts of the Nature Conservancy and the influence of private individuals. One of these was the aviator and conservationist, Charles A. Lindbergh, whose gravesite lies in the Kipahulu district close to the park boundary.

Kipahulu remains in pristine condition primarily because of its inaccessibility. It is an area of virgin forests where the Maui nukupuu, thought to be extinct, was rediscovered and where other rare birds have ecological niches.

As most new maps of Hawaii reflect the effects of development on the islands in the form of residential areas, industrial sites, roads and structures, one appreciates more and more the wisdom of early conservationists and lawmakers in protecting and safeguarding areas like our national parks. Although changes have occurred in these areas as the maps show, since becoming national parks, they remain for the most part untouched by man.

The terrestrial dynamism that has made the Kilauea, Mauna Loa and Haleakala areas so unique is still very active. As the monitoring, study, and presentation of the data to illustrate the dynamics of the park areas continue, maps, imagery, photos and other graphics are developed. Even the newest maps by the United States Geological Survey, Bier, and Nelles became obsolete in the summer of 1989 when lavas from the Southeast Rift Zone overran the Chain of Craters road again and covered the Wahaula Visitors Center and Heiau. In 1990 lava over-ran the village and black sand beach at Kalapana. This is not the end by any means. As the national parks in Hawaii continue to grow and change by acts of God, and occasionally Congress, maps and other spatial documents will record the change and development of this very special and unique corner of the globe.

NOTES

*This paper is based on a presentation given at the American Library Association Annual Conference, Dallas, Texas, June 1989.

1. Gavan Daws, *Hawaii: the Islands of Life* (Honolulu: Signature Publishing for the Nature Conservancy of Hawaii, 1988), 17.
2. *Volcanism in Hawaii,* U.S. Geological Survey Professional Paper 1350 (Reston, Virginia: The Survey, 1987), 287, 519, 574.
3. Gary Fitzpatrick and Riley Moffat, *Palapala'aina: the Early Mapping of Hawaii* (Honolulu: Editions Limited, 1986), 97.
4. Charles H. Hitchcock, *Hawaii and Its Volcanoes* (Honolulu: The Hawaiian Gazette Co., 1909), 193.

5. Isabella L. Bird, *Six Months in the Sandwich Islands,* 1875 reprint (Honolulu: University of Hawaii Press for Friends of the Library of Hawaii, 1964), 201.
6. William T. Brigham, *The Volcanoes of Kilauea and Mauna Loa on the Island of Hawaii,* Bishop Museum Memoirs, vol. 2, no. 4 (Honolulu: The Museum, 1909), 123.
7. Brigham, *Volcanoes of Kilauea and Mauna Loa,* 158-161.
8. Hitchcock, *Hawaii,* 260.
9. Daws, *Hawaii,* 17.
10. *Island of Hawaii, Territory of Hawaii, Map and Guide* (Honolulu: Hawaii Promotion Committee, 1910).
11. Bird, *Six Months,* 53.
12. Harold T. Stearns, *Memoirs of a Geologist* (Honolulu: Hawaii Institute of Geophysics, 1983).
13. U.S. National Park Service, *Draft Land Protection Plan, June 1985* (Honolulu: The Service, 1985).
14. Louise Armitage and George T. Armitage, *Self-Guiding Map of Hawaii National Park* (Honolulu: Hawaii Natural History Association, 1949).
15. *Designating a Portion of the Hawaii National Park as the Haleakala National Park,* 86th Congress, 2nd Session. Senate Report 1655 (Serial 12236). House Report 2170 (Serial 12248).

SELECTED CARTOBIBLIOGRAPHY

Alexander, William DeWitt. *Map of the Crater of Haleakala* [map]. 1:70,000. London: Cooper & Hodson, 1869. In: Bird, Isabella L. *Six Months in the Sandwich Islands.* 1875 reprint. Honolulu: University of Hawaii Press, 1966. p. 200.

Alexander, William DeWitt. *Map of the Hawaiian Islands* [map]. 1:360,000. Philadelphia: J. L. Smith, 1876.

Armitage, Louise and George T. Armitage. *Self-Guiding Map of Hawaii National Park* [map]. Scales vary. Honolulu: Hawaii Natural History Association, 1949.

Baker, Joseph. *A Chart of the Sandwich Islands*... [map]. 1:860,000. In: Vancouver, George. *A Voyage of Discovery to the North Pacific Ocean*. London: G. G. and J. Robinson, 1798.

Barringer, W. H. and Thomas Boles. *Hawaii* [map]. 1:365,000. Hilo: The Automobile Club of Hawaii, 1923.

Bier, James A. *Map of Hawaii* [map]. 4th ed. 1:250,000. Honolulu: University of Hawaii Press, 1988.

Bier, James A. *Map of Maui* [map]. 4th ed. 1:154,000. Honolulu: University of Hawaii Press, 1988.

Brigham, William T. *The Crater of Kilauea in 1865* [map]. 1:38,600. In: Dutton, Clarence Edward. "Hawaiian Volcanoes," *Fourth Annual Report of the United States Geological Survey*. Washington: GPO, 1884, opposite p. 118.

Dodge, Frank S. *The Crater of Kilauea, Hawaii* [map]. 1:26,000. In: Brigham, William T. *The Volcanoes of Kilauea and Mauna Loa on the Island of Hawaii*. Honolulu: Bishop Museum Press, 1909. p. 163.

Dodge, Frank S. *Maui, Hawaiian Islands* [map]. 1:60,000. New York: Julius Bien, 1885.

Dodge, Frank S. "Plans and Sections of Halemaumau Showing Changes in Crater Since March 1886" [ms. map]. 1:6,000. Hawaiian Government Survey Reg. No. 1829.

Donn, John M. *Hawaii, Hawaiian Islands* [map]. 1:190,080. Honolulu: Hawaii Territorial Survey, 1901.

Hawaii Territorial Survey. *Map of the Island of Hawaii* [map]. 1:240,000. Honolulu: Hawaii Territorial Survey, 1928.

Malden, Charles R. *A Plan of the Volcano Peli in the Island of Owhyhee* [map]. 1:47,500. In: Byron, Lord George Anson. *Voyage of H.M.S. Blonde to the Sandwich Islands*. London: John Murray, 1826.

Mann, S. E. "Map of the Great Crater of Haleakala" [ms. map]. 1:15,840. April 1884. Bishop Museum.

Map of Part of the Island of Hawaii [map]. 1:191,300. In: U.S. Exploring Expedition. *Atlas of Charts*. Philadelphia: C. Sherman & Son, 1858, plate 62.

Newton, H. R. *The Hawaiian Islands* [map]. 1:720,000. Honolulu: Hawaii Territorial Survey, 1919.

Roberts, Henry. *Chart of the Sandwich Island* [map]. 1:1,500,000. In: Cook, James. *A Voyage to the Pacific Ocean . . .* London: W. & A. Strahan, 1784.

Stearns, Harold T. and Gordon A. MacDonald. *Geological and Topographic Map of the Island of Hawaii* [map]. 1:125,000. In: Stearns and MacDonald *Geology and Grand-water Resources of the Island of Hawaii*. Bulletin 9. Honolulu: Hawaii Division of Hydrography, 1845.

Stearns, Harold T. *Island of Maui, Hawaii: Geologic and Topographic Map* [map]. 1:62,500. In: Stearns and MacDonald. *Geology and Ground-water Resources of the Island of Maui, Hawaii*. Bulletin 7. Honolulu: Hawaii Division of Hydrography, 1942.

U.S. Geological Survey. *Hawaii National Park (Kilauea-Mauna Loa Section) Island of Hawaii* [map]. 1:62,500. Washington: U.S. Geological Survey, 1927.

U.S. Geological Survey. *Hawaii Volcanoes National Park and Vicinity, Hawaii* [map]. 1:100,000. Reston: U.S. Geological Survey, 1986.

U.S. Geological Survey. *Proposed Kilauea Volcano National Park* [map]. 1:31,680. Washington: U.S. Geological Survey, 1912.

U.S. Geological Survey. *Topographic Map of the Island of Maui, Maui Country, Hawaii* [map]. 1:62,500. Washington: U.S. Geological Survey, 1931.

U.S. Geological Survey. *Topographic Map of the Island of Maui, Maui Country, Hawaii* [map]. 1:62,500. Washington: U.S. Geological Survey, 1961.

U.S. National Park Service. *Topographic Map of the Island of Hawaii, County of Hawaii* [map]. 1:125,000. Honolulu: Hawaii Division of Hydrography, 1944.

Wall, William A. *Map of the Island of Hawaii, Hawaiian Islands* [map]. 1:240,000. New York: Julius Bien, 1887.

ISLAND WILDERNESS: ISLE ROYALE NATIONAL PARK

Nancy J. Edstrom

Conservation Laboratory

Illinois State Historical Library

When looking at a map of Lake Superior, it takes little imagination to see its resemblance to the head of a large wolf. The neck is truncated by the shoreline stretching from Marquette, Michigan, to the United States-Canadian border at Sault Sainte Marie. The nose lies at Duluth, Minnesota; the ears at Nipigon, Ontario; and Isle Royale is its eye.[1]

Isle Royale National Park is small by National Park Service standards. It is an archipelago consisting of one large island, 45 miles long and 9 miles wide at its broadest point, surrounded by over 200 smaller islands of greatly varying size. Park boundaries extend 4.5 miles into Lake Superior, making the total park area some 571,000 acres of which approximately one-quarter is land. Isle Royale is located in the state of Michigan and is its northernmost point, although it is actually much closer to Ontario and Minnesota. From the Greenstone Ridge, the island's backbone, it is possible to see Thunder Bay and the Sleeping Giant in Sibley Provincial Park on the southern coast of Canada and even a tiny part of Minnesota.[2]

Isle Royale roughly parallels the northern shore of Lake Superior and is oriented in a southwest-northeast direction. It is characterized by a series of ridges and valleys which follow the island's orientation. They were formed in Precambrian times when lava seeped into the Superior Basin and formed the region's basalt bedrock. After each flow, sediments were carried into the area ,producing soft rock between the basalt layers. The center of the basin subsided, raising and tilting the layers of rock which became Isle Royale, the Keweenaw Peninsula, and much of the outer rim of the lake. Erosion of the soft rock formed the valleys, and the layers of basalt became the ridges which continue underwater when they reach Lake Superior, emerging to form the narrow islands. Glaciers scoured the land and left smooth, rounded rock and debris which formed thin layers of soil sufficient to support the growth of trees. Open strips of grass and shrubs stretch along the tops of some of the ridges through which the underlying rock mass is often visible.[3]

Retreating glaciers also caused the formation of Isle Royale's inland lakes, forty-two of which are named. The largest of these, Siskiwit Lake, contains its own islands. A favorite superlative of those familiar with Isle Royale concerns Ryan Island, which is located in Siskiwit Lake. It is the largest island in the largest lake on the largest island in the largest freshwater lake in the world.[4]

Isle Royale's insular nature makes it unique within the National Park System. No other park in the forty-eight contiguous states has the qualities of both wilderness and isolation to such a marked extent.[5] The remoteness which inspired the movement to make it a national park is what draws visitors to the island. This same remoteness, however, makes Isle Royale one of the least-visited national parks in the United States. Approximately 12,000 to 15,000 people visit the park each year; the same number visit Yellowstone on one weekend. One factor which affects the number of visitors is the length of the visitor season. Isle Royale is the only national park to close completely from November 1 to April 15 each year. Even the park headquarters moves, from Mott Island, one of the small surrounding islands, to Houghton, in Michigan's Upper Peninsula. Another factor is the relative difficulty of access and the need for special transportation. Visitors generally travel to the island by boat, although a seaplane also makes regular trips. In addition, travel can be held up due to the unpredictability of Lake Superior. A third factor is the park's facilities. While some visitors choose to stay at the Rock Harbor Lodge or the housekeeping units, many are backpackers who hike the island's 170 miles of trails or canoe the inland lakes. These last two factors also affect the length of time visitors stay at Isle Royale. In 1983 the average time spent per person for all other national parks was four and a half hours; at Isle Royale, it was four days.[6]

Isle Royale's insular nature has also made it a valuable site for scientific study. It is the single most studied area in the state of Michigan.[7] The island's plant and animal communities have been both protected from outside influences by Lake Superior and affected by the isolation from other species. Three of the inland lakes have been separated from each other long enough for new and distinct species of fish to have developed, and the Isle Royale red squirrel is also a unique sub-species.[8] The most well-known and longest ongoing study focuses on the predator-prey relationship between the island's wolf and moose populations. This study began in 1958 and continues today. Although only a few species of mammals are known to live on the island, it maintains natural food chains representative of boreal regions and is unique in that top carnivores, such as the wolf, still exist and are naturally regulated.[9] The park's forests also reflect the effect of its insularity. White spruce, balsam fir, paper birch and aspen, classic boreal forest types, ring the shores nurtured by the cold of Lake Superior. A hardwood forest of sugar maple and yellow birch grows on warmer inland ridges. In recognition of its unique habitats, Isle Royale was named an International Biosphere Reserve as part of UNESCO's Man and the Biosphere Program in 1981.[10]

NATIVE AMERICAN AND EARLY EUROPEAN MAPPING

Human activity on Isle Royale dates back some 4,000 years. By about 2000 B.C. Indians were mining copper on the island. Wood from one pit has been radiocarbon-dated at 2130 B.C., plus or minus 130 years. This early mining activity lasted 1,000 years.[11] Archaeological records are sketchy at best from this point until about 300 B.C. At that time the Indians frequented Isle Royale rather extensively, although they did little mining. Based on archaeological studies, the period from 800 to 1600 saw the peak of Indian activity on the island. After Europeans arrived in the Great Lakes area, Indian culture on Isle Royale began to disintegrate and their numbers declined.[12]

The Indians made maps cut in birchbark or wood, or drawn on paper for European explorers. These maps were more often concerned with mythological and religious associations of areas than with accurate geographical representation. The birchbark scrolls made by the southern Ojibway of Minnesota are an example. They show the migration route followed by the Ojibway from their ancestral home on the shores of the Atlantic to their nineteenth-century home in northern Minnesota.[13] They are vague towards the right, or east, in their representation of the area of origin and early migration. Topographic detail becomes increasingly recognizable towards the left, or west, where they illustrate the later and more clearly recollected migrations through the use of definite pictographic clues. This is especially true for the area between the head of Lake Superior and Leech Lake, Minnesota.[14]

Although it cannot be proven, it is possible that the first European to visit Isle Royale was Étienne Brulé. Brulé was a young Frenchman sent by Samuel de Champlain to live among the Huron Indians. It is believed that, in 1623, he traveled through the Georgian Bay to the Sault Sainte Marie Rapids at the foot of Lake Superior.[15] Although no record of his voyage exists, it is theorized that he passed above the Sault Sainte Marie and coasted the northern shore of Lake Superior to the mouth of the Saint Louis River near present-day Duluth. It is also assumed that he visited Isle Royale on the return trip.[16] This assumption is based on Champlain having received knowledge of the existence of a mine of "pure copper" on an island.[17]

Pierre Esprit Radisson was the first European to give proof of a visit to Lake Superior[18] and provided the first account of Isle Royale.[19] In 1659 he and Medard Chouart, Sieur de Groseilliers, set out for the Sault Sainte Marie by way of the Ottawa River. After reaching Lake Superior, they proceeded along the southern shore until they reached the Keweenaw Peninsula which they crossed through Portage Lake. While crossing near the present sites of Houghton and Hancock, the Indians told them that

> when they are minded to thwart it [Lake Superior] in a faire and calme wether, beginning from sun rising to sett, they come to a great island [Isle Royale], from whence they come the next morning to firme lande att the other side.

Radisson, however, did not visit Isle Royale himself.[20]

The first accurate description of the Lake Superior region was compiled by the French Jesuit missionaries. The narrative is included in a compendium of letters known as the *Relations* which were written by the Jesuits to their supporters in Europe. The *Relation* of 1669-1671, edited by Father Claude Dablon, includes the following description of Isle Royale:

> But farther toward the West, on the same North side, is found the Island which is most famous for Copper, and is called Minong; this is the one in which, as the Savages have told many people, the metal exists in abundance, and in many places. It is large, and is fully twenty-five leagues long; it is distant seven leagues from the mainland, and more than sixty from the end of the lake. Pieces of

> Copper, mingled with the stones, are found at the water's edge almost all around the Island.[21]

Dablon's description is not limited to the single large island. He also wrote: "This large Island is almost all surrounded with Islets that are said to be formed of Copper; they are encountered in various places, as far as the mainland on the North."[22]

The *Relation* for 1670-1672 contains the first European map of Lake Superior. Father Dablon noted that it was the work of "two priests," probably Dablon himself and Father Claude Allouez who, in 1665, travelled along the southern coast of Lake Superior to Chequamegon Bay where he established a mission. Two years later Allouez set out for Lake Nipigon and, after arriving on the northern shore of Lake Superior, proceeded northeast until he reached Isle Royale. He then went on to Lake Nipigon.[23] The accuracy of the Jesuit map, amazing considering the size and nature of Lake Superior and the available instrumentation, is apparent when compared with modern maps of the lake. The Keweenaw and Bayfield peninsulas are correctly located east and west (although too far south), and the width of the lake at those points is almost exactly correct. The relative accuracy of the location of the islands is notable.[24] Isle Royale is accurately portrayed, including the islands at the mouth of Washington Harbor and Passage Island northeast of the main island. This implies that the Jesuit fathers were familiar with the archipelago.[25]

It would seem that the Jesuit map should have become the model for the delineation of Lake Superior on maps of North America but this was seldom the case. During the years following its production, cartographic depictions of Lake Superior in general, and Isle Royale in particular, became increasingly distorted. The location and shape of the island and the size of the lake were misrepresented.[26] In 1744 Jacques-Nicolas Bellin published his map *Carte des lacs du Canada* which contained the most glaring error in the mapping of Lake Superior: the introduction into print of several imaginary islands. While some of them seem to actually exist and were given different names, two of them have never been adequately explained. Bellin was a trained engineer with a substantial reputation for being scrupulous about his sources. His maps were often accompanied by printed

Remarques that listed and critiqued existing maps. He wrote six pages on Lake Superior in his long essay on his North American maps but said nothing about the appearance of the islands, with the exception of listing their names. Both Bellin and Father Pierre-Francois-Xavier de Charlevoix, whose works the map was commissioned to accompany, have been blamed for the appearance of the islands. Recent studies may have uncovered their true source.[27]

The archives of the Départment de la Marine in Paris holds two manuscript maps which were probably the basis for Bellin's map. One is inscribed "fait a quebec le 25 Septembre 1735 Chaussegros de Léry fils" and the other, though anonymous, can be attributed to his father, Chaussegros de Léry (père), who was the chief engineer and Bellin's counterpart in New France. His son, who began training in surveying and cartography in 1733 at the age of twelve, was serving as apprentice draftsman to his father. These maps seem to have been made as the result of the travels of Louis Denys de La Ronde, an officer in the navy and colonial troops, and were probably based on information which he provided. One of these manuscripts shows all of Lake Superior, and all the islands included on Bellin's map are shown in the same general positions. In addition, all the islands seem to have been named for those people who were in positions to help La Ronde, who hoped to receive a nine-year monopoly on the fur trade at Chagouamigon free of licensing fees in exchange for his building two ships to find and exploit the Lake Superior copper deposits.[28] It is not known whether the imaginary islands were a deliberate hoax or simply the result of misinterpreting Indian accounts. While La Ronde never claimed to have seen the islands, his somewhat shady character casts doubt upon his intentions.[29] In any event, the mythical islands on Bellin's map seem to have come from these manuscript maps. In his *Remarques*, Bellin indicated that he had used manuscripts located in the Department de la Marine and wrote:

> Lake Superior, the largest of the lakes known in America, is poorly represented on all maps and one can see at first glance how much I have changed it. Documents from the Dépot have allowed me to represent it more faithfully than we have seen it until now.[30]

Bellin may have concluded that the manuscripts were the better maps because they were newer than the Jesuit map. Once in print, and bolstered by his reputation, the imaginary islands continued to appear on maps of that region for the next century.[31]

THE TREATY OF PARIS

Isle Royale's inclusion in the United States is a direct result of Bellin's error. John Mitchell began making a map of North America in 1750 and published it in London five years later. This map has been called the most important map in American history.[32] Mitchell's map was made under orders of the Lords of Trade and Plantations to show as English territory the extent and features of a country hotly contested by England and France. At the time the Lake Superior country had not been fully explored, and there was little knowledge of its geography. Bellin's map served as the basis for Mitchell's delineation of Lake Superior.[33]

Even thought it was considered "very defective and extremely erroneous,"[34] Mitchell's map was used in negotiations with Great Britain which resulted in the Treaty of Paris in 1783. This treaty ended the Revolutionary War and recognized the independence of the United States. According to a popular legend, Benjamin Franklin, one of the negotiators, knew about the copper deposits found on Isle Royale and secured it for the United States. This has since been disproven.[35] Rather, the negotiators wanted to construct a water boundary that would be easily identified. The British were given two boundary choices by the Americans: the 45th parallel extended to the Mississippi River; or through the Great Lakes waterways to the northwest corner of the Lake of the Woods, then following the 49th parallel to the Mississippi. The British chose the latter.[36] On the Mitchell map, the Mississippi appeared to extend quite far north, although an inset map covered its upper course. It was not known at the time that the source of the Mississippi was located nowhere near the 49th parallel. In addition, Isle Royale was placed much farther south than it actually is. Based on assumptions made from looking at Mitchell's map, the northern boundary of the United States was placed ". . . northward of the isles, Royal and Philippeaux . . ." [one of the imaginary islands].[37] Many years passed before the boundary was finally settled. The present boundary was not formally agreed upon until the Webster-Ashburton Treaty in 1842.[38] Because Isle Royale was named specifically in the treaty, subsequent surveys had to retain it for the United States. As a result, the United States received a larger share of Lake Superior than did British North America.[39]

European exploration of the Lake Superior region continued during the eighteenth and nineteenth centuries, but the journeys generally took place along the northern and southern shores of the lake. Fur traders developed a major trade route along the lake's northern shore. At Grand Portage, Minnesota, merchants built a fort, warehouses, and a supply depot. From this base, traders radiated to all parts of the North American continent during the eighteenth and nineteenth centuries. Isle Royale, only twenty miles from Grand Portage, lay largely unexplored.[40]

From the eighteenth until the mid-nineteenth century, descriptions of Isle Royale were made by explorers who never set foot on the island. Jonathan Carver, a Connecticut-born soldier, was in the western Lake Superior region in 1766. Although he expressed interest in the lake's islands, he did not make an effort to visit them. His account of Isle Royale is even less exact than that of the Jesuits, made one hundred years earlier:[41]

> There are many islands in this lake, two of which are very large; and if the land of them is proper for cultivation, there appears to be sufficient to form on each a considerable province; especially on Isle Royale, which cannot be less than an [*sic*] hundred miles long, and in many places forty broad. But there is no way at present of ascertaining the exact length or breadth of either.[42]

Major Stephen Long led an expedition through the northern territories of the United States in 1823. Long traveled along the northern shore of Lake Superior. He did not stop to investigate Isle Royale but did write in his journal: "We have had Isle Royal[e] in view yesterday and today. It is the largest of the islands of Lake Superior . . . its surface is somewhat broken, rising into elevated hills in some parts and presenting swells in others."[43]

In 1823 Lieutenant Henry Wolsey Bayfield, a British naval officer, was sent by the Admiralty to accurately chart the Superior basin. He spent three years on the lake, circumnavigating its shores and surveying its islands and the rivers flowing into it. In addition he did some preliminary field work on Isle Royale itself and published the first geological report on the island and a flattering description of its copper resources.[44] In 1825 Bayfield returned to England where he spent the next two years preparing charts of the Great Lakes, including Lake Superior, in the Hydrographic Office of the Admiralty, London. The charts of Superior were completed in 1827.[45] His chart of the north shore includes a map of Isle Royale which identifies, for the first time, the sheltered anchorages and hidden inlets. Some twenty years after its publication, a visitor to Isle Royale reported that:

> The only correct chart of the entire coast [of Lake Superior] is that made by the British Government . . . under the able superintendence of Capt. Bayfield . . . and to this hour, the only map used by the Ordnance department at Washington and by the superintendent of the mineral lands at Porter's Island, is that of Bayfield.[46,47]

MICHIGAN STATEHOOD

On January 26, 1837, Michigan became the 26th state to enter the Union. At the time, the entire Lake Superior region was inhabited by Chippewa Indians who were considered a dependent domestic nation and who had legal title to the lands which they lived upon. The Indians' rights interfered with the exploration of the region's mineral resources, so federal government negotiators met with Chippewa leaders to discuss limiting actual Indian title. The last of a series of treaties, which began in 1836, was the Treaty of La Pointe, signed in 1842. It gave the federal government control of all lands from east of present-day Marquette, Michigan, north to the United States-Canadian border, and west to the Saint Louis River, including all the Lake Superior islands. The treaty seemed to clearly include Isle Royale, but a misunderstanding by negotiators created some confusion. The Chippewa of Grand Portage, the band most closely associated with the island, were not invited to take part in the treaty negotiations because they were thought to be living under British jurisdiction and, therefore, had no claim to land within the United States. In addition, it was the position of the federal government negotiators that because all of the leaders of the Chippewa had signed the treaty Isle Royale had been included and that one band could not hold out because all of the land belonged to the entire tribe. The Chippewa, as well as hundreds of whites and mixed bloods who attended the negotiations, felt that Isle Royale could not have been included in the treaty since the Grand Portage band had not been present.[48] They proposed to sell Isle Royale for $60,000 but, after a year and a half of negotiating, they signed the Isle Royale Compact in 1844, turning over their rights to the island for $400 worth of gunpowder and $100 worth of beef.[49]

At the same time negotiations were being conducted with the Chippewa, planning continued for both exploration and exploitation the Lake Superior mineral lands. Little was known about the interior of the Lower Peninsula at the time of Michigan's admission to the Union, and even less was known about the region farther north. As a result, Michigan's new legislature, just two hours after Michigan achieved statehood, introduced and passed a bill which proposed and funded a thorough geological study of the state.[50] The driving force behind this geological study was Douglass Houghton, Michigan's first state geologist. He conceived and planned it, as well as persuaded members of the legislature to commit time and money to it.

After an initial survey in 1837, a second bill that enlarged the survey was passed in March 1838. It provided additional funding that made provision for zoological, botanical, and topographical departments as well as the geological department. However, the financial problems that affected the entire country at that time took their toll and, within a year, the departments of zoology and botany had been suspended. The geographical and topographical surveys continued, with the topographer providing Houghton and his assistants with "skeleton" plats of townships based on the federal linear survey at the scale of two inches to the mile. The geologists were to fill in geological, topographical, and civil features and return them to the topographer. The topographer would reduce them and send them to be engraved and printed. It was through this process that a series of new and extremely accurate and detailed maps of the state and counties were to be produced. Due to the poor financial condition of the state, however, only five of the county maps were ever published.

Douglass Houghton and his assistants began explorations in the Upper Peninsula of Michigan in 1839 and finally reached Isle Royale in 1840. In his fourth annual report to the Michigan legislature, dated February 11, 1841, Houghton remarks:

> The published maps of that portion of the state of Michigan usually known as the upper peninsula, are so defective, not only in details, but also in general outline, that the task of giving a description, of any portion, in such a manner as to render it intelligible to any person who has not actually traveled over it, is exceedingly difficult. The extent of these geographical inaccuracies is much greater than would at first be supposed, for scarcely a single feature of the interior is given as it actually exists; . . . . even the coast maps of the great lakes, by which our upper peninsula is in part surrounded, are usually so defective as scarcely to be recognized, except in their most general outlines.[51]

He does, however, praise the map of the coast of Lake Superior published by the Society for the Diffusion of Useful Knowledge. This map was based on the surveys of Henry Bayfield, and Houghton considered the American coastline "faithfully delineated in its general outline but in minutiae it is frequently deficient."[52] Houghton's report also included a general description of Isle Royale and a discussion of its mineral reserves, along with those of the rest of the Upper Peninsula. Although he cautioned against creating unfounded expectations about the region's mineral resources, he concluded that the area would provide a source of steadily increasing wealth to the state.[53]

By the end of 1842, Houghton and his assistants had completed most of their fieldwork and had begun to organize the large amount of geological and topographical data they had collected. They also continued to fill in the skeleton township plats. In his annual report dated February 15, 1844, Houghton describes similar activities for 1843. He also discusses a plan for completing the survey of the Upper Peninsula at little cost to the state. Houghton was not entirely satisfied with the work that he and his assistants had done in the Upper Peninsula. He felt that it was necessary to survey it in much more detail than the Lower Peninsula. The state could not afford to finance such a detailed survey, so Houghton proposed to link it to the federal linear survey, having the deputy surveyors make geological, topographical, and mineralogical observations while carrying out the business of surveying the townships and ranges. Houghton presented his plan to the Secretary of the Interior and to Lucius Lyon, the United States Surveyor General. Lyon approved of the plan, but doubts were raised about whether deputy surveyors who had the appropriate scientific background to undertake such work with any degree of success could be found. Houghton himself offered to take a contract, an action which helped to convince the federal government to adopt the plan. He signed a contract with the commissioner of the General Land Office on June 25, 1844, to complete the survey of the Upper Peninsula.[54]

Houghton arranged to subcontract some of the work for the survey. One of those whom he hired was William Burt, who had first contracted for public land surveys in Michigan in 1833. Burt was the inventor of the solar compass, an instrument necessary because of the effect of Michigan's iron ore deposits on the needles of magnetic compasses. Burt and Houghton began surveying in August 1844. One member of their party was William Ives, who served as compass man. Ives also made surveys for Houghton in 1845. Over forty townships had been surveyed by the fall of 1845, and the exploration of the Upper Peninsula was almost completed when tragedy struck. Houghton and two other men were drowned in Lake Superior off the coast of the Keweenaw Peninsula.[55]

The Michigan Geological Survey was suspended after Houghton's death. However, Bela Hubbard, one of Houghton's assistants, continued to survey for the General Land Office, as did William Burt and William Ives. One of the last parts of the state to be surveyed was Isle Royale, which Ives signed a contract to do on April 23, 1847. In the contract, he agreed to make a complete survey, including topographical features, geological formations, and references to mines and minerals. He signed a handwritten agreement with Bela Hubbard on May 1, 1847, agreeing to collect specimens and notes for Hubbard whom he hired to serve as the geologist for the survey.[56]

Ives and a party of six men left Detroit for Isle Royale on May 14, 1847, and arrived on May 21st. They began surveying the next day and worked until September 28th; of that time, 118 days were actually spent surveying. The physical nature of the island made the work more difficult. Because of its orientation, it does not contain one complete township. In addition, noting topographical features took a great deal of time. Ives spent much of his time meandering—surveying the banks of streams or the area of lakes. In a journal Ives kept while on Isle Royale, he mentions meandering 98 of the 118 days of surveying during 1847.[57] The survey proceeded without incident until July 30th when Ives received a letter from Bela Hubbard saying that he had resigned as geologist. Dr. Charles Jackson, who had been appointed United States geologist for Michigan's mineral lands by Congress in March 1847, had taken over the geological part of the survey. On August 10th, Ives received another letter, this time from Lucius Lyon explaining that the contract they had signed for the survey of Isle Royale was invalid with regard to the geological aspect. This also invalidated Ives' contract with Hubbard. Lyon offered Ives a new contract for the land survey only, which Ives accepted.[58]

By September 28th, Ives' survey was partially completed. He and his men left the island for the winter and returned the next May to finish. They met with Charles Jackson's assistant geologist on July 14th and furnished him with plats of the island and other information which would be useful to him. Ives and his men continued onward and finally completed their survey on July 20th. They proceeded to La Point, where Ives prepared plat maps and five field books of notes for the Surveyor General's office. Ives then left to survey other parts of the Great Lakes states.[59]

Charles Jackson's geological survey continued after the land survey was finished. He spent little time at Isle Royale since he was responsible for surveying all of northern Michigan, and what time he did spend there was concentrated near the shore. He depended on his assistants to explore the interior of the island. Assistant geologist George J. Dickenson traveled with William Ives during 1847, after Jackson took over the geological survey from Bela Hubbard, and returned again the following year. Two others who assisted Jackson were John W. Foster and Josiah D. Whitney. Foster and Whitney continued Jackson's work after the latter resigned from the survey after only one season's work. In 1849 Jackson wrote a report on the Michigan mineral lands consisting of his own notes and material gathered by Douglass Houghton before his death. The report, issued as both a House and a Senate executive document of the 31st Congress, 1st session, was massive and non-analytical and represented an optimistic view of the region's mineral potential. Included among its plates and maps are two very similar geological maps of Isle Royale. One of these was done by Jackson and the other by Foster and Whitney.[60]

Foster and Whitney also wrote a report on their geological examination of the Lake Superior mineral region. Their report, more sober and scientific than Jackson's, was issued in two parts. The first part, concerning the copper lands, was issued as a House Executive Document of the 31st Congress, 1st session, in 1850. The second part, on the iron region with a general geological description of the whole area, was a Senate Executive Document from a special session of the 32nd Congress in March 1851. Although it contained many illustrations, only four maps were included, none of which was the Isle Royale map that had been present in Jackson's report. Foster and Whitney incorporated the work of many others into their report, including Douglass Houghton, Charles Jackson, and George Dickenson. This information, along with their own field observations, provided a detailed description of Isle Royale and the surrounding mineral region. They were concerned with other aspects of the region as well, such as its botany and archaeology, in addition to its economic geology. While they were optimistic about the mineral potential of the mainland, they did not feel the same way about Isle Royale.[61]

Interest in Isle Royale was not limited to that of scientists. By the time Jackson's and Foster and Whitney's reports were published, commercial copper mining operations had been in place on the island for several years. From 1843 to 1881, Isle Royale mines contributed 2.8 percent of Michigan's copper product.[62] Commercial fishing in Isle Royale waters proved to be even more profitable. Beginning in the 1830s, within a few years it surpassed all other areas in Lake Superior in fish production. It remained an important part of island commerce well into the 1980s. With this economic growth came a parallel growth in island population. Such population changes, along with those of the entire Upper Peninsula, brought about changes in political jurisdiction.

Initially, the land in the Upper Peninsula ceded by the Indians in 1842 was organized into a single large county called Chippewa. The following year Ontonagon County was created, composed of the far western quarter of the Upper Peninsula and Isle Royale. In 1845 Isle Royale was moved again, this time into Houghton County which consisted of territory west of Ontonagon County.[63] In March 1861 Keweenaw County was created from the northern tip of Houghton County and included Isle Royale. The island continued under this jurisdiction until March 1875 when it was established as an independent county. Initially it consisted of one township with the county seat at the town of Island Mine. Later, it was divided into two townships, and the county seat moved to the town of Minong. Isle Royale County proved to be short-lived, however. In 1885 its judicial functions were assumed by Houghton County. The county came to an end on April 9, 1897, when it was eliminated and absorbed by Keweenaw County.[64]

It may be that Isle Royale County's brief history as an independent entity accounts for the apparent scarcity of maps showing it as such, and little has been written about it. Many contemporary maps of the region do not show the island at all, such as the map of Michigan and Wisconsin appearing in *Mitchell's New General Atlas.* The county is included, however, in a table of county populations; Isle Royale County is listed as having 55 inhabitants in 1880.[65] This table also appears on a map of Michigan prepared for the state commissioner of immigration that is included in *Michigan and Its Resources,* published in 1881. Isle Royale appears in an inset and the locations of Island Mine and Minong are indicated (fig. 16).[66] The author has been able to find only one map clearly indicating Isle Royale as an independent county. It appears as an inset on a map of Michigan and Wisconsin drawn by A. von Steinwehr and published in 1877 by Van Antwerp, Bragg & Co.

TOURISM AND PARK STATUS

As time passed, commercial interest in Isle Royale gradually waned. Commercial fishing continued at a smaller scale, and the last of the mining ventures ended in 1893. Much of the reason for this decline lay

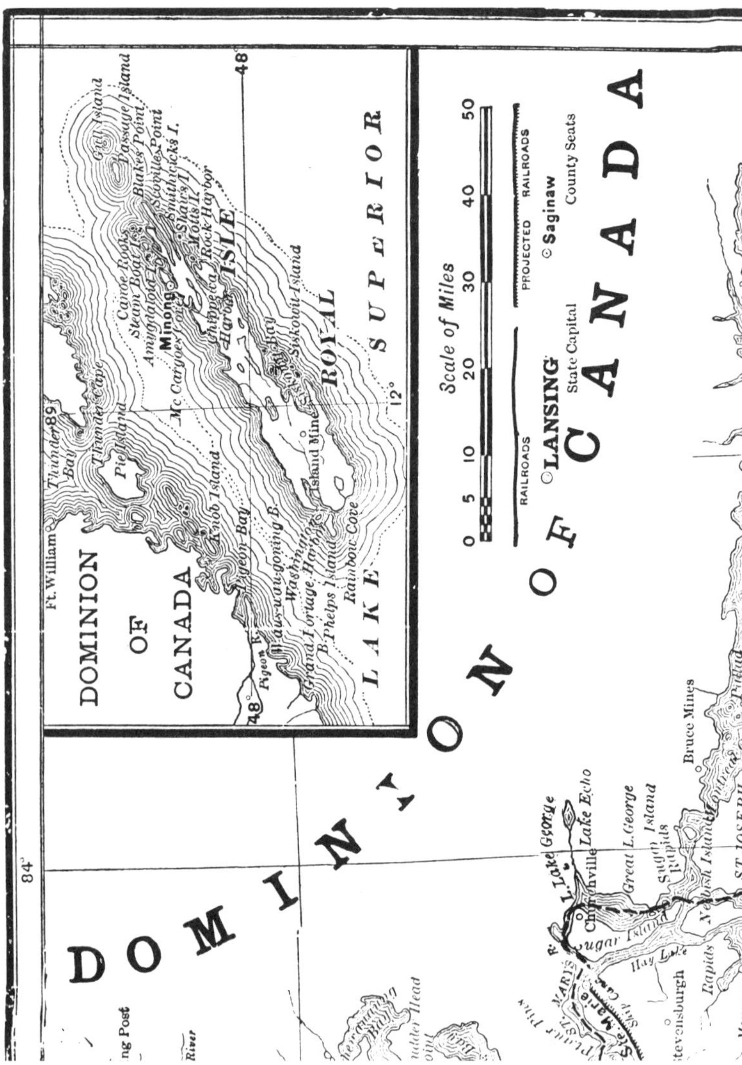

Fig. 16. Inset of Ilse Royale showing the towns of Island Mine and Minong from Michigan by Henry S. Stebbins, 1881. (From the Map Library, University of Michigan)

in the island's remoteness and the great difficulty of transporting equipment and products between it and the mainland. This very same remoteness, however, proved to be an asset to yet another industry: tourism. Largely free from the pollution and pollen of more populated areas, Isle Royale had been a welcome haven of relief for those suffering from hay fever since at least the 1860s.[67] It was also visited by groups of Michigan and Minnesota sport fishermen who were drawn by the abundant fish population. By 1920 there were several resorts on the island. One had a nine-hole golf course and a swimming pool the owner created by walling off a section of Lake Superior that the sun warmed. The most important resort was one located at Rock Harbor and owned by Kneut Neutson, a Minnesota businessman. He first visited in 1898 and fell in love with the island. He bought a block of land around Rock Harbor and platted out streets and house lots, after which he built a resort he called Park Place (fig. 17). His resort was the basis for what is the present-day Rock Harbor Lodge.[68]

Its increasing popularity as a recreational area called attention to the need to preserve Isle Royale in its natural state. After a visit to the island in the fall of 1921, Albert Stoll, outdoor editor of the *Detroit News*, began a campaign to achieve national park status for Isle Royale. The campaign gained momentum throughout the 1920s as first the state and then the federal government withdrew their land holdings from sale in 1922 and 1924 respectively. As it progressed, the campaign reawakened the scientific community's interest in the island, culminating in a project funded by the Michigan legislature and carried out by the University of Michigan.[69]

In 1929 a bill was passed authorizing $15,000 to finance a thorough study of Isle Royale by the University of Michigan.[70] The study emphasized four areas: anthropology, botany, geography and geology, and zoology.[71] The geographical team was charged with the responsibility of making an overall, systematic geographic survey with a special emphasis upon vegetational conditions. The result was a report which was accompanied by a large-scale map of vegetation associations, neither of which was ever published. This map was compiled not only from traditional ground surveys but also from a series of aerial photographs taken through a cooperative arrangement with the United States Geological Survey.[72]

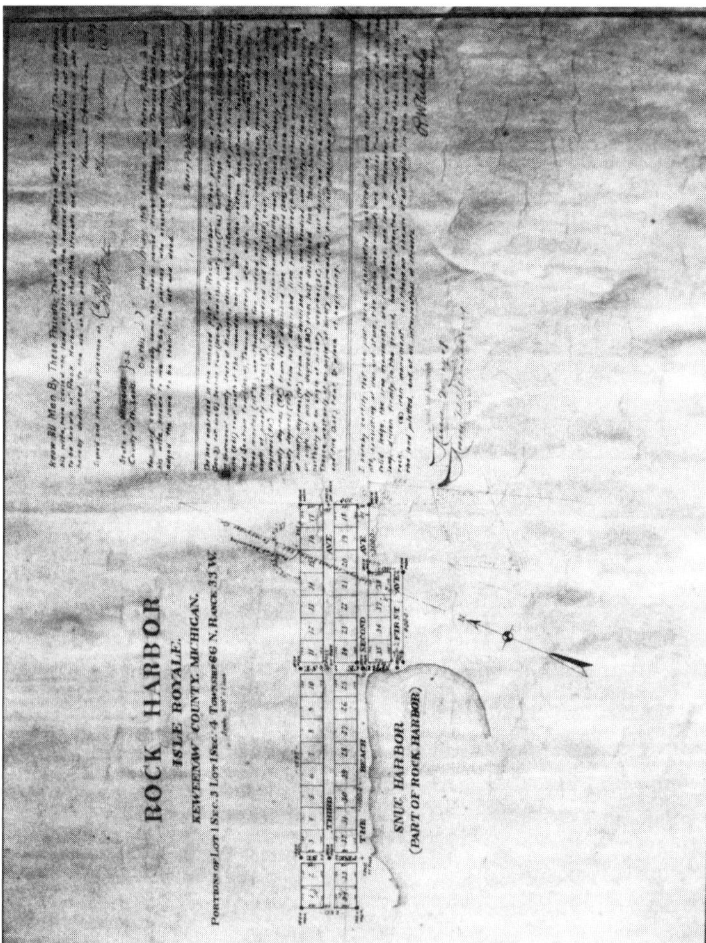

Fig. 17. Deed and plat of Rock Harbor issued to Knuet and Theresa Neutson on May 27, 1908. (From the collections of Isle Royale National Park)

The aerial photographs of Isle Royale were taken by Talbert Abrams, the founder of the Abrams Aerial Survey Corporation of Lansing, Michigan. Abrams' task was a difficult one due to Isle Royale's remote location. He and his pilot could not use a hydroplane because the pontoons would block the camera's view, so once their airplane left the closest airport at Houghton, Michigan, some 60 miles away, there would be no safe place for it to land until it returned to Houghton. Abrams received final authorization to proceed with the aerial survey on July 25, 1930, and actually began the job on August 6th. Although he and his pilot made 16 round trips to the island and spent a total of 60 hours in the air, traveling some 6,000 miles, it took until September 3rd for Abrams to finish the job. He encountered a variety of problems: mechanical difficulties with cameras and the plane, inclement weather, film ruined by static electricity in the air, and five days of delay due to forest fires. The fires generated enough smoke to cover the lake to a height of 10,000 feet, and it was too smoky over the island to map it.[73] When he finished, Abrams had produced a set of 680 photographs at a scale of 1:12,000 (fig. 18). In addition to the ground cover map, the photographs were used by the United States Geological Survey to compile a base map of Isle Royale which was issued in March 1931.[74]

Bills that would enable Isle Royale to become a national park were introduced into the United States House and Senate in February 1931. They passed with only minor discussion, and on March 3rd a joint version was signed into law.[75] The federal enabling legislation stated that when all remaining land had been donated to the federal government and when the state of Michigan had ceded jurisdiction, then the area would become a national park. It also stipulated that no federal funds could be used to purchase land and that it could only be secured by public or private donation.[76]

Meanwhile, the Michigan Department of Conservation offered its Isle Royale holdings to the federal government. The Michigan legislature created a five-man Isle Royale National Park Commission to assist in negotiating the land acquisition.[77] It would, however, take almost ten years to complete the process. The Great Depression slowed the momentum that the park movement had gained. As a result, the commission languished and by 1934 only one member

Fig. 18. Aerial photograph of Isle Royale showing Chickenbone Lake and Lake Livermore from the aerial survey by Talber Abrams, August 1930. (From the Map Library, University of Michigan)

remained.[78] The threat of timber-cutting on the island rekindled public interest in the park and the commission was reorganized. It resumed the process of obtaining title to Isle Royale land.[79] The federal government assisted by establishing Civilian Conservation Corps (CCC) camps on the island and providing money to purchase land from the National Industrial Recovery Act. Land acquisition continued and finally, on April 3, 1940, the National Park Service declared that Isle Royale was officially a national park.[80]

The physical planning for the national park had begun in 1935. In August of that year, the first group of CCC enrollees arrived on the island, including the men who would create the "master plans" for the proposed park. The text of a "master plan" sets forth the scheme for the development of an area. It describes in detail such matters as nature and location of proposed structures, roads and trails, landscaping and reforestation, signs, or any other modifications. The text is accompanied by a series of maps showing in detail the location of the proposed modifications and their relationship to the rest of the area. Typically, it includes a topographic map, road system and trail plans, a fire-control map, a vegetation type map, administrative and utility areas plans, and other miscellaneous plans.[81] The first edition of the master plan, dated January 1936, was a text-only document setting forth the principal park values and proposed guidelines to be followed in the preparation of future master plans. The second edition (January 1937) included a general development plan and a land ownership map. The third edition (January 1938) was the first edition to include a fire protection plan. This was a high priority as a result of a fire that burned nearly 20% of the island during the summer of 1936. Also included were plans for Rock Harbor Lodge and the park's summer headquarters on Mott Island.[82]

The CCC played a major role in the development of the park. When the master plans were being created, projects were also planned for CCC enrollees. They constructed trails across the island, managed wildlife, built park facilities for visitors and Park Service staff, and cleaned up fire-damaged areas.

Although the park was established in 1940, the dedication did not take place until August 27, 1946, because of the occurrence of World War II.[83] During the years 1941-1945, visitor attendance declined sharply. There was little sign of improvement after the war due to the condition of transportation to and from the island. Several of the national parks had fallen into a state of disrepair, and they soon were the object of criticism. Conrad L. Wirth, who became director of the Park Service in 1951, used this criticism to his advantage, proposing a ten-year development program that would require major financial allocations. This program, known as Mission 66, was supported by Congress and the president. Over a billion dollars were spent on improvements throughout the whole park system, with Isle Royale's share coming to more than $3 million. This money was used to upgrade visitor facilities and to purchase a new boat, the *Ranger III*, for ferrying visitors to the park. These improvements helped to stimulate park attendance and the number of visitors increased.[84]

With the new surge of interest in Isle Royale, the need for new maps to supplement the park brochures became evident. Consequently, in 1955 the first tourist map was issued, drawn by Robert M. Linn. It has since been updated periodically and is now issued by the Isle Royale Natural History Association, a non-profit organization which assists the park by publishing and selling educational materials. Isle Royale's points of interest are not limited to those on land. The island is surrounded by a number of shipwrecks, ten of which have been included in a group nomination to the National Register of Historic Places. Small numbers of sport divers began to explore the wrecks during the 1960s. Now several hundred visit each year.[85] In order to preserve and protect the shipwrecks, the National Park Service's Submerged Cultural Resources Unit has studied and mapped them. As an experiment, the *Monarch*, which lies off the northern tip of Isle Royale, was marked with a "trail" that was to have been described in a waterproof interpretive guide. However, the feedback from divers was largely negative and the plan was abandoned.[86]

The first detailed topographic maps of the island were not issued until 1958. A 15-minute topographic sheet covering Passage Island at the northeast end of Isle Royale had been issued in 1895 but nothing else was produced for over sixty years. The new sheets, which were part of the United States Geological Survey's regular topographic series, were drawn at a scale of 1:62,500 and published in two editions. In one edition, relief was indicated by both shading and contours; in the other, it was shown by contours alone. These maps were revised in 1970. Topographic maps at a scale of 1:24,000 were not published until 1985. This series consists of 14 sheets covering the entire park. A similar series of planimetric maps was issued the same year. These are at a scale of 1:100,000. Finally, a single-sheet map at a scale of 1:62,500 was published in 1987. It is printed on waterproof material, making it suitable for hikers and backpackers to use in their treks.

Throughout the history of Isle Royale cartography, mapmakers have been confronted with a variety of special challenges. The very remoteness of the island caused it to be largely ignored until the 1840s. What maps existed before that date were crude and often inaccurate. It was not until the time of Douglass Houghton and William Ives that reliable maps based on firsthand observation became available. In the following years, the incentive for mapmaking lay largely in scientific and, more recently, recreational interest. Although scientific study continues, the future of the mapping of Isle Royale seems likely to emphasize the latter. With these maps in hand, visitors will continue to be better able to appreciate and enjoy the natural beauty of this island wilderness.

NOTES

*This paper is based on a presentation given at the American Library Association Annual Conference, Dallas, Texas, June 1989.

1. William Ashworth, *The Late Great Lakes: An Environmental History* (New York: Alfred A. Knopf, 1986), 226.
2. Ibid., 226-27.
3. Jim DuFresne, *Isle Royale National Park: Foot Trails and Water Routes* (Seattle: The Mountaineers, 1984), 16.
4. U.S. Department of the Interior, National Park Service, *Isle Royale National Park, Michigan* (Washington, D.C.: Government Printing Office, 1986), [1].
5. Theodore J. Karamanski and Richard Zeitlin, with Joseph DeRose, *Narrative History of Isle Royale National Park* (Chicago:

Mid-American Research Center, Loyola University of Chicago, 1988), 312.

6. John L. Eliot, "Isle Royale: A North Woods Park Primeval," *National Geographic* 167 (April 1985): 540.

7. Karamanski, Zeitlin, and DeRose, *Narrative History*, 3.

8. Michigan Technological University, Department of Biological Sciences, *Isle Royale Biosphere Reserve: A History of Scientific Studies* (Atlanta: Science Publications Office, National Park Service, 1985), 53-54.

9. Ibid., 49-50.

10. Eliot, "Isle Royale," 540.

11. Napier Shelton, *The Life of Isle Royale* (Washington, D.C.: Office of Publications, National Park Service, 1975), 21-22.

12. Ibid.

13. Robert W. Karrow, Jr., "Lake Superior's Mythic Isles: A Cautionary Tale for Users of Old Maps," *Michigan History* 69 (January/February 1985): 25.

14. G. Malcolm Lewis, "The Indigenous Maps and Mapping of North American Indians," *The Map Collector* no. 9 (December 1979): 26.

15. Karamanski, Zeitlin, and DeRose, *Narrative History*, 24.

16. Nellis M. Crouse, *Contributions of the Canadian Jesuits to the Geographical Knowledge of New France, 1632-1675* (Ithaca, N.Y.: Cornell University, 1924), 126.

17. Willshire Butterfield, *History of Brulé's Discoveries and Explorations, 1610-1626* (Cleveland: Helman-Taylor Company, 1898), 159.

18. Crouse, *Contributions of the Canadian Jesuits*, 128.

19. Karamanski, Zeitlin, and DeRose, *Narrative History*, 23-24.

20. James B. Griffin, ed., *Lake Superior Copper and the Indians: Miscellaneous Studies of Great Lakes Prehistory*, University of Michigan Museum of Anthropology, Anthropological Papers, no. 17 (Ann Arbor: University of Michigan, 1961), 38.

21. Reuben Gold Thwaites, ed., *The Jesuit Relations and Allied Documents: Travels and Explorations of the Jesuit Missionaries in New France 1610-1791*, vol. 54 (New York: Pageant Book Company, 1959), 159.
22. Ibid., 161.
23. Crouse, *Contributions of the Canadian Jesuits*, 132-133.
24. Karrow, "Lake Superior's Mythic Isles," 27.
25. Karamanski, Zeitlin, and DeRose, *Narrative History*, 26.
26. Ibid., 29.
27. Conrad E. Heidenreich, "The Fictitious Islands of Lake Superior," *The Map Collector* no. 27 (June 1984): 21-22.
28. Ibid., 22-24.
29. Karrow, "Lake Superior's Mythic Isles," 29.
30. Heidenreich, "Fictitious Islands," 25.
31. Karrow, "Lake Superior's Mythic Isles," 30.
32. *Dictionary of American Biography*, vol. 13 (New York: Charles Scribner's Sons, 1934), 50-51.
33. Karrow, "Lake Superior's Mythic Isles," 30.
34. Anthony Barclay, quoted in ibid.
35. James C. Starbuck, "Ben Franklin and Isle Royale," *Michigan History* 46 (June 1962): 166.
36. D. Robert Hakala, *Isle Royale: Primeval Prince* [rev. ed.] (Washington, D.C.: National Park Service, 1955), 18.
37. Griffin, *Lake Superior Copper*, 30-31.
38. Karrow, "Lake Superior's Mythic Isles," 31.
39. Karamanski, Zeitlin, and DeRose, *Narrative History*, 30.
40. Ibid., 27-28.
41. Ibid., 28.
42. Jonathan Carver, *Travels through the Interior Parts of North-America, in the Years 1766, 1767, and 1768* (London: J. Walter and S. Crowder, 1778), 134.

43. Major Stephen Long, quoted in Karamanski, Zeitlin, and DeRose, *Narrative History*, 32-33.

44. Ruth McKenzie, ed., *The St. Lawrence Survey Journals of Captain Henry Wolsey Bayfield, 1829-1853*, vol. I (Toronto: The Champlain Society, 1984), xxiii-xxiv.

45. Ibid., xxv.

46. Public Archives of Canada, Film A-424, "From Correspondence of Livingston & Co.'s Express, No. 10 Wall Street, Sault Ste. Marie, July 12, 1845," extracted from the New York *Commercial Advertiser* and enclosed in letter, Bayfield to Beaufort, 26 Nov. 1845, rpt. in ibid., xxvi.

47. The next navigational chart of Isle Royale was not published until 1871 when one was produced by the Army Corps of Engineers. The engineers of the Survey of the Northern and Northwestern Lakes reached the island in 1867 and executed surveys every year through 1871. The chart was extremely valuable to scientists since it provided them with a base map that could be used in their reports. Karamanski, Zeitlin, and DeRose, *Narrative History*, 353; Hakala, *Isle Royale: Primeval Prince*, 26.

48. Karamanski, Zeitlin, and DeRose, *Narrative History*, 39-40.

49. Russell M. Magnaghi, "The Isle Royale Compact of 1844," *Inland Seas* 33 (Winter 1977): 289-291.

50. Arthur A. Socolow, ed., *The State Geological Surveys: A History* (N.p.: Association of American State Geologists, 1988), 223.

51. George N. Fuller, ed., *Geological Reports of Douglass Houghton, First State Geologist of Michigan, 1837-1845* (Lansing, Mich.: The Michigan Historical Commission, 1928), 484.

52. Fuller, *Geological Reports*, 484n.

53. Ibid., 191.

54. Alvah Bradish, *Memoir of Douglass Houghton, First State Geologist of Michigan* (Detroit: Raynor & Taylor, 1889), 39-41.

55. C. Albert White, *A History of the Rectangular Survey System* (Washington, D.C.: U.S. Department of the Interior, Bureau of Land Management, [1983]), 101-102.

56. Marie E. Gilchrist, "Isle Royale Survey, Part I," *Inland Seas* 24 (Fall 1968): 184.

57. Ibid., 191.

58. Marie E. Gilchrist, "Isle Royale Survey, Part II," *Inland Seas* 24 (Winter 1968): 318-20.

59. Marie E. Gilchrist, "Isle Royale Survey, Part III," *Inland Seas* 25 (Spring 1969): 37-39.

60. Page 625 of the report includes a message from Foster and Whitney dated July 25, 1849, stating that "In the course of a month, or six weeks, we will transmit a geological map of Isle Royale, constructed on similar principles, which will complete the geology of all the 'copper region.' "

61. Karamanski, Zeitlin, and DeRose, *Narrative History*, 45-50.

62. During the mid to late 1800s, approximately 80 percent of the nation's copper came from Michigan mines.

63. In 1847 Isle Royale and its adjacent islands were established as a township.

64. Hakala, *Isle Royale: Primeval Prince*, 62-63; John H. Long, ed., *Historical Atlas and Chronology of County Boundaries, 1788-1980* (Boston: G. K. Hall, 1984), vol. 3, *Michigan, Wisconsin*, 74-75, 82, 88-89, 120-21.

65. S. Augustus Mitchell, *Mitchell's New General Atlas, Containing Maps of the Various Countries of the World, Plans of Cities, etc. Embraced in 93 Quarto Maps* (Philadelphia: Bradley & Co., 1883), 31.

66. Throughout its history, the name of the island was variously spelled as "Isle Royale" and "Isle Royal." In the second report of the U.S. Board on Geographic Names for 1890-1899, the official spelling was listed as "Isle Royal" (U.S. Board on Geographic Names, *Second Report*, 2nd ed. (Washington: Government Printing Office, 1901), 72.). However some three decades later, bowing to local practice, the board reinstated spelling the name as "Isle Royale" (*Decisions of the United States Geographic Board* no. 14 (Feb. 3, 1932), 2.).

67. John J. Little, "Island Wilderness: A History of Isle Royale National Park," (Ph.D. diss., University of Toledo, 1978), 15.
68. Ibid., 15-16; Karamanski, Zeitlin, and DeRose, *Narrative History*, 219-220.
69. Martha M. Bigelow, "Isle Royale National Park Movement or a Study in Frustrations," *Michigan History* 41 (March 1957): 37.
70. *The Compiled Laws of the State of Michigan, 1948*, Vol. 2, Act 209, 1929 (Ann Arbor, Michigan: The Ann Arbor Press, 1948), 6237.
71. Bigelow, "Isle Royale National Park Movement," 38.
72. Kenneth C. McMurray, interview by Martha M. Bigelow, December 12, 1955, transcript, Michigan Historical Collections, Bentley Historical Library, University of Michigan, Ann Arbor, Mich.
73. Albert Stoll, Jr. and Talbert Abrams, "13 Thrilling Minutes in a Storm Cloud: Mapping Isle Royale from Plane, Fliers Cheat Death by Inches," *The Detroit News*, January 11, 1931, Feature-Fiction Section.
74. Kenneth C. McMurray, "Isle Royale Survey, University of Michigan, Geographical Report," (Department of Geography, University of Michigan, [1930], Typescript), 38.
75. Little, "Island Wilderness," 37-40.
76. Statutes at Large, 1514.
77. Little, "Island Wilderness," 41.
78. Ibid., 45.
79. Karamanski, Zeitlin, and DeRose, *Narrative History*, 324-325.
80. Ibid., 335.
81. "Preliminary Checklist of the Records of the National Park Service: Master Plans of Areas Administered by the National Park Service," (Division of Interior Archives, September 15, 1942, Typescript).
82. Donald Wolbrink, "Isle Royale Recollections, 1935-1936-1937," (Summer 1987, Typescript), 10-12.

83. Donna L. Root, "The Dedication of Isle Royale National Park," *Inland Seas* 2 (October 1946): 269.

84. Little, "Island Wilderness," 157-159; Karamanski, Zeitlin, and DeRose, *Narrative History*, 336-338.

85. Eliot, "Isle Royale," 544.

86. Mary Maruca, "Treasures of the Deep," *National Parks* 56 (March/April 1982): 15-16; Bruce Webber, letter to author, August 10, 1990.

SELECTED CARTOBIBLIOGRAPHY

Abrams Aerial Survey Corporation. *Aerial Photographic Survey Map, Isle Royale, in Lake Superior, Keweenaw County, Michigan, August 1930* [aerial photographs]. 1:12,000. Lansing, Michigan: Abrams Aerial Survey Corporation, 1930.

Bayfield, Henry Wolsey. *Lake Superior, Chart II* [map]. Ca. 1:343,000. Survey of Lake Superior. London: Hydrographic Office of the Admiralty, 1828.

Bayfield, Henry Wolsey. *Chart of Part of the North Coast of Lake Superior, from Grand Portage Bay to Hawk Island, including Isle Royale* [map]. Ca. 1:865,000. London: Hydrographic Office of the Admiralty, 5th May 1828.

Bellin, Jacques-Nicolas. *Cartes des lacs du Canada* [map]. Ca. 1:3,150,000. Paris, 1744.

Chaussegros de Léry, Gaspard-Joseph, 1721-1797. ["East Shore of Lake Superior from Saulte Sainte Marie to the Michipicoton River"] [ms. map]. 8"=20 lieues. September 25, 1735. Bibliothèque, Service Hydrographique, Paris.

Chaussegros de Léry, Gaspard-Joseph, 1682-1756. "Carte du Lac Superieur" [ms. map]. 9 3/8"=40 lieues. [1735?]. Bibliothèque, Service Hydrographique, Paris.

Foster, John W. and Josiah D. Whitney. *Geological Map of Isle Royale* [map]. Ca. 1:129,969. New York: Ackerman, Lith., [1849]. In: U.S. Senate, 31st Cong., 1st Sess. *Report on the Geological and*

Mineralogical Survey of the Mineral Lands of the United States in the State of Michigan, made under the authority of an act of Congress approved March 1, 1847... (S. Ex. Doc. 1, Pt. 3). Washington, D.C.: Printed for the Senate, 1849. (Serial Set No. 551)

[Isle Royal] [inset map]. Ca. 1:1,450,000. On: Stebbins, Henry S. *Michigan* [map]. Ca. 1:1,450,000. Chicago: Henry S. Stebbins, October 1881. In: Michigan. Commissioner of Immigration. *Michigan and its Resources.* Lansing, Michigan: W. S. George & Co., 1881.

Isle Royale Co., Michigan [inset map]. 1:2,343,000. On: von Steinwehr, A. *Map of Michigan and Wisconsin* [map]. 1:2,343,000. Cincinnati and New York: Van Antwerp, Bragg & Co., 1877. In: *Special Geography of the North Central States East of the Mississippi.* Cincinnati and New York: Van Antwerp, Bragg & Co., 1878.

Ives, William. "Land Survey Notes and Plats of Isle Royale, Michigan" [ms. maps]. 1:31,680. 1848. Files of U.S. Bureau of Land Management, U.S. Department of the Interior, Washington, D.C.

Jackson, Charles T. *Geological Map of Isle Royale* [map]. Ca. 1:142,783. New York: Ackerman, Lith., [1849]. In: U.S. House, 31st Cong., 1st Sess. *Report on the Geological and Mineralogical Survey of the Mineral Lands of the United States in the State of Michigan, made under the authority of an act of Congress approved March 1, 1847...* (H. Ex. Doc. 5, Pt. 3). Washington, D.C.: Printed for the Ho. of Reps., 1849. (Serial Set No. 571)

Also in: U.S. Senate, 31st Cong., 1st Sess. *Report on the Geological and Mineralogical Survey of the Mineral Lands of the United States in the State of Michigan, made under the authority of an act of Congress approved March 1, 1847...* (S. Ex. Doc. 1, Pt. 3). Washington, D.C.: Printed for the Senate, 1849. (Serial Set No. 551)

Lac Superieur et autres lieux ou sont les missions des Peres de la Compagnie de Iesus comprises sous le nom D'outaouacs [map]. 5"=50 lieues. In: *Relation de ce qui s'est passe de plus remarquable aux missions des Peres de la Compagnie de Iesus en la Nouvelle France, les annees 1670 & 1671.* Paris: Chez Sabastien Mabre Cramoisy, 1672.

McMurray, Kenneth C. *[Isle Royale Forest Cover Type]* [map]. Ca. 1:12,800. [Ann Arbor: Department of Geography, University of Michigan, 1932].

Mitchell, John. *North America with the Roads, Distances, Limits, and Extent of the Settlements, humbly inscribed to the Right Honourable the Earl of Halifax, and the other Right Honourable the Lords Commissioners for Trade & Plantations* [map]. Ca. 1:2,000,000. London: Andrew Millar, 1755.

Nichols, Ray W. "Rock Harbor, Isle Royale, Keweenaw County, Michigan" [ms. map]. 1:2,400. [1908]. On: "Original Deed for Town of Rock Harbor, Issued to Kneut and Theresa Neutson on May 27, 1908." Park Headquarters, Mott Island, Isle Royale National Park.

Raynolds, William F. *Chart of Isle Royale, including North West Coast of Lake Superior* [map]. 1:120,000. Washington, D.C.: Corps of Engineers, War Department, [1871?].

Society for the Diffusion of Useful Knowledge. *Lake Superior* [map]. Ca. 1:1,800,000. London: Chapman and Hall, 1832.

U.S. Geological Survey. *7.5-Minute Series (Topographic) [maps]. 1:24,000. Reston, Va.: U.S. Geological Survey, 19- .*

Belle Harbor (1985)
Feldtmann Lake (1985)
Feldtmann Ridge (1985)
Lake Richie (1985)
Little Todd Harbor (1985)
Malone Bay (1985)
McCargoe Cove (1985)
Mott Island (1985)
Passage Island (1985)
Point Houghton (1985)
Rock Harbor Lodge (1985)
Sugar Mountain (1985)
Todd Harbor (1985)
Windigo (1985)

U.S. Geological Survey. *1:100,000-Scale Planimetric Map* [maps]. Reston, Va.: Geological Survey, 198 - .

Grand Portage, Minnesota-Michigan (1985)
Pigeon Point, Minnesota-Michigan (1985)
Siskiwit Bay, Michigan (1985)
Todd Harbor, Michigan (1985)

U.S. Geological Survey. *Isle Royale, Keweenaw County, Michigan* [map]. 1:31,680. Washington, D.C.: U.S. Geological Survey, 1931.

U.S. Geological Survey. *Isle Royale National Park: 1:62,500-scale topographic map* [map]. 1:62,500. Reston, Va.: U.S. Geological Survey, 1987.

U.S. Geological Survey. *Topographic Map of Isle Royale National Park, Keweenaw County, Michigan* [map]. 1:62,500. Washington, D.C.: U.S. Geological Survey, 1958.

U.S. National Park Service. *The Master Plan, Isle Royale National Park* [maps]. 2nd edition. Scales vary. [Washington, D.C.: National Park Service], 1937.

U.S. National Park Service. *The Master Plan, Isle Royale National Park* [maps]. 3rd edition. Scales vary. [Washington, D.C.: National Park Service], 1938.

YELLOWSTONE PARK: THE FIRST OF MANY, THE GRANDEST OF ALL

Jim Walsh

Thomas P. O'Neill, Jr. Library
Boston College

. . . though small in extent, I

regard the valley of the upper Yellowstone as the most interesting unexplored district in our widely expanded country.[1]

We saw many strange and wonderful phenomena, many things which would require volumes for adequate description, and which in future geography will be classed among the wonders of the earth.[2]

We may say, in conclusion, that we have been able . . . to do little more than allude to a few of the wonderful physical phenomena of this marvelous valley.[3]

Capt. William F. Raynolds', Lt. Gustavus C. Doane's, and Dr. Ferdinand V. Hayden's descriptions of the Yellowstone country were some of the most detailed and accurate of their day and are still considered to be among the best narratives of the region and its features. Their reports, accompanying maps, illustrations, and photographs were instrumental in obtaining Congressional approval for the establishment of Yellowstone National Park. President Ulysses S. Grant signed the Yellowstone National Park Act on March 1, 1872, and the world's first national park was established.[4] The idea of setting aside a parcel of land that would preserve forever the unique and majestic features of the Yellowstone country was first suggested by Thomas Meagher, the acting territorial governor of Montana, in 1865.[5]

The amount of material written about Yellowstone National Park and its exploration is voluminous, and it would be impossible to cover every aspect of every expedition that explored and mapped the park in a paper of this length.[6] Major expeditions and maps will be discussed and compared, while minor events will be mentioned or referenced. The time period of 1803 to the present will be covered; however, this paper will concentrate on two specific time periods: 1859-1871 and 1872-1879. The expeditions that occurred, the discoveries that were made, the narratives that were written, and the reports that were published were the impetus for the establishing the park and preserving it for the future. These events and activities served to inform and educate the American people about the spectacular features which existed in what we now know as Yellowstone National Park.

EARLY YEARS, 1803-1859

The United States purchased the Louisiana Territory from France on April 11, 1803, and with it began the exploration of America's West. President Thomas Jefferson realized the importance of exploring and mapping the territory. He authorized the Lewis and Clark Expedition (1804-1806) to explore the newly acquired territory. Captains Meriwether Lewis and William Clark's party, known as the Corps of Discovery, left St. Louis on May 14, 1804, and embarked on the first United States government sponsored expedition.[7] Although their discoveries were monumental and their expedition set precedents for all future expeditions, they never entered the present day boundaries of Yellowstone National Park.

John Colter was a member of the Corps of Discovery and asked to leave the expedition early. Permission was granted and in August 1806 he set out on his own and remained in the mountains. He spent the winter of 1807-1808 in the Yellowstone country and is acknowledged as being the first white man to enter and view the wonders of what is now Yellowstone National Park.[8]

Colter returned to St. Louis in May 1810 and provided valuable information on what he saw to William Clark. Clark used this information on his 1806-1811 manuscript map and Samuel Lewis included it on his engraving of Clark's map, which was printed in 1814.[9]

These two maps were the first to include features of the Yellowstone country, albeit limited and incomplete. Eustis Lake (now Yellowstone Lake) and Lake Biddle, referred to as Lake Riddle on Samuel Lewis' map (now Jackson Lake), are notable features on the two maps. The relative size and location of the lakes are somewhat accurate on the 1806-1811 map, but they are inaccurately portrayed as being nearly equal in size on Lewis' version.[10]

The exploration of the West continued and peaked between the 1830s and 1850s. These expeditions were concerned with the exploration of the entire West and not just the microcosm of the Yellowstone country. None of these expeditions entered the present day boundaries of the park, and as a result, the information which appeared on their maps was lacking in detail and often inaccurate. The information was second hand, provided by the trappers, hunters, prospectors, and mountain men who had visited the region. Colter, Jim Bridger, Joseph Meek, and William Sublette returned from their visits with stories of geysers, bubbling mud pots, and petrified forests. Everyone believed that these accounts were only the creation of their vivid and colorful imaginations. Most people did not believe that such features or phenomena existed. This skepticism may have been one reason why people did not attempt to settle in this area, thus leaving the region intact and pristine.

Nevertheless, some of trappers' information did find its way onto some of the most important maps of the time. These included Washington Hood's map of 1839, John Frémont's map of 1848, Father Pierre De Smet's map of 1851, and G. K. Warren's map of 1857.[11] However, these maps consisted of large portions of "white area," i.e., unexplored area, and they added no new information to the mapping of the Yellowstone country.

THE GOLDEN AGE OF EXPLORATION, 1859-1872

1859 marked the beginning of a thirteen-year period of heightened interest, exploration, and mapping of the Yellowstone country. The expeditions of this period and their subsequent reports, maps, illustrations, and, eventually, photographs were the catalyst for the establishment of Yellowstone National Park.

Raynolds Expedition, 1859-1860

Captain William F. Raynolds led a Corps of Topographical Engineers expedition on the first organized expedition whose orders included the exploration of the Yellowstone country.[12] The exploring party, which included Jim Bridger as guide and Ferdinand Hayden as naturalist, left St. Louis in May 1859 and camped that winter in Wyoming.[13] In May 1860, at the junction of the Popo Agie and Wind rivers, the expedition was divided into two parties. Lieutenant Henry E. Maynadier led one party around the eastern and northern boundaries of the park, and Captain Raynolds led the other party around the southern and western boundaries of the park.[14] Together, both parties completely encircled the park but never crossed the present day boundaries.

Raynolds had planned to traverse the Wind River Mountains, diagonally cross the park, and meet Maynadier at Three Forks. If his plan had been successful he would have been the first person to enter the Yellowstone and officially report on the wonders found there. He failed to cross the mountains due to the snow pack. He could not wait for the snows to melt because his orders also required him to be in the British Territory by July 18, 1860, to view a total eclipse. In his official report, which was not published until 1868 due to the outbreak of the Civil War, Raynolds wrote of his failure to enter Yellowstone:

> Had our attempts to enter this district been made a month later in the season, the snow would have mainly disappeared, and there would have been no insurmountable obstacles to overcome. I cannot doubt, therefore, that at no very distant day the mysteries of this region will be fully revealed . . .[15]

Raynolds' expedition was significant for many reasons, in spite of his failure to enter the present day boundaries of Yellowstone National Park. It was the first expedition, as mentioned previously, whose orders included the Yellowstone country as a goal. It was the last expedition organized under the auspices of the Corps of Topographical Engineers.[16] It served as a transition from one generation of explorers (Frémont, Kearny, and Warren) to another (Hayden, King, and Powell). His report and accompanying map, which was published separately in 1864, provided, for the time, accurate and up-to-date information for an unknown part of the nation. It was said that

> Raynolds' report was the first official recognition in any form of the probable existence of extensive volcanic phenomena in the region of the Upper Yellowstone.[17]

Raynolds' map, *U.S. War Department Map of the Yellowstone and Missouri Rivers and Their Tributaries Explored by Captain W. F. Raynolds Topl. Engrs.* includes both his and Maynadier's routes and depicts various features of the Upper Yellowstone Valley, as described by Jim Bridger to Captain Raynolds (fig. 19). Also, Sulpher Mt. is in the

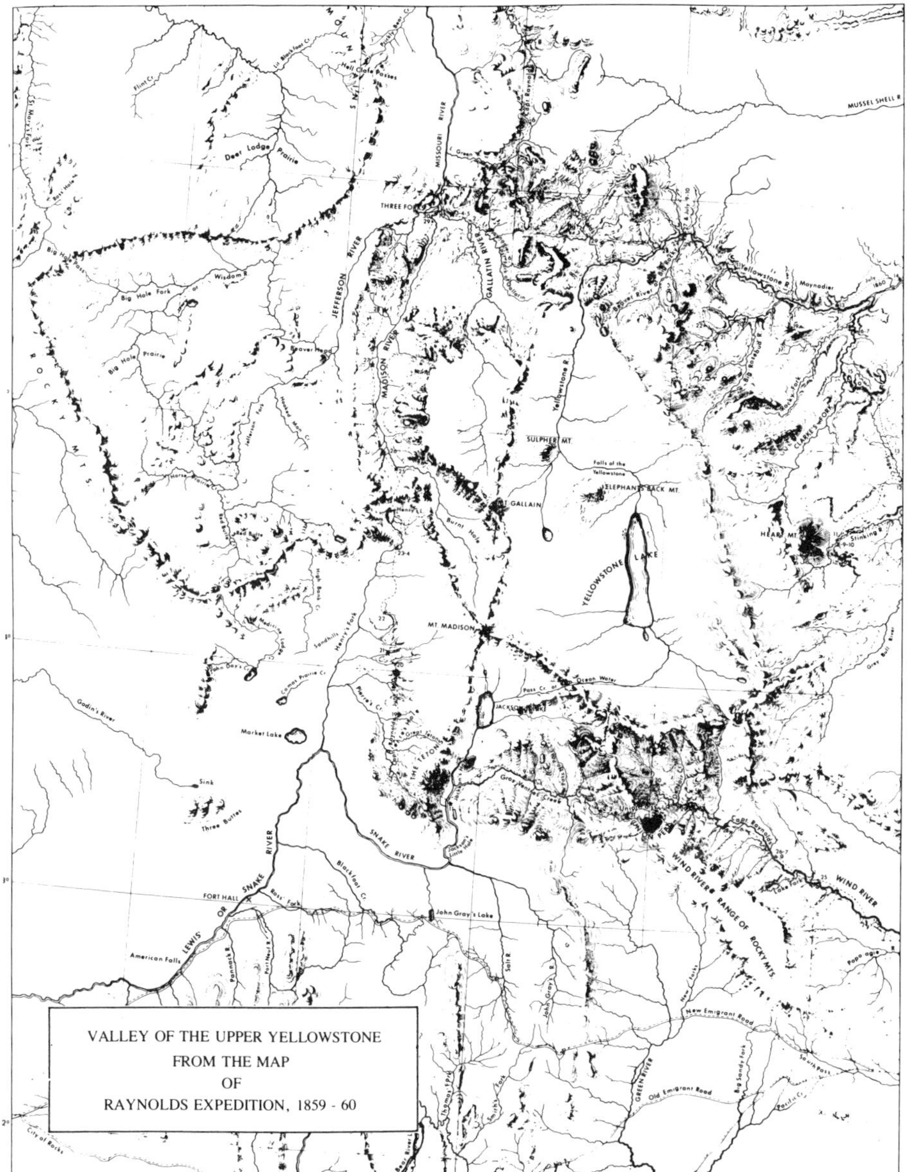

Fig. 19. From Baldwin's *Enchanted Enclosure,* 1976.

correct location of the Mammoth Hot Springs and Mt. Gallatin appears in the position of present day Mount Holmes.[18] However, many of the features are shown too far north and west (e.g., part of Jackson Lake and all of the Tetons are located in present day Idaho, not Wyoming).

It is unfortunate that Raynolds' report was not published until 1868 which resulted in it being less timely and an "account [that] was largely stale news."[19]

deLacy Prospecting Trip, 1863

In 1863 Walter deLacy led a prospecting party which passed through the southwestern corner of present day Yellowstone National Park. This was the first organized expedition to enter the present day boundaries of Yellowstone National Park, although their primary purpose was prospecting not exploring the Yellowstone country. He failed to publish his discoveries and did not receive the credit which he rightly deserved.[20]

In 1864 deLacy was commissioned by the Montana Territory Legislature to compile a map of the territory, and in 1865 *Map of the Territory of Montana with Portions of the Adjoining Territories* was published. This map contained a few of the features of the Yellowstone country (Yellowstone Lake was similar in shape and location to Raynolds' Yellowstone Lake). The map was reprinted, improved, and corrected during the next 24 years.[21] The 1870 edition of this map, hereinafter referred to as the *deLacy Map of 1870*, which deLacy compiled with David Folsom, was the first map that showed many of Yellowstone's features and showed them with a high degree of accuracy.[22]

Folsom-Cook Expedition, 1869

In 1867 and 1868, expeditions were planned to explore the Yellowstone country but sporadic Indian uprisings and the lack of adequate military escorts caused them to be cancelled. David Folsom, Charles Cook, and William Peterson departed from Diamond City,

Montana, on September 6, 1869, in spite of these "inconveniences." As a result, they became the first organized expedition which set out to explore the Yellowstone country and accomplished its objective.[23]

They traveled up the Yellowstone River, through the Grand Canyon of the Yellowstone, and observed and measured the Lower and Upper falls. They then preceded along the northwest shore of Yellowstone Lake, skirted the northernmost point of Shoshone Lake, traveled north through the Lower Geyser Basin, past the Great Fountain Geyser, and exited the park following the Madison River. They returned to Helena, Montana, on October 11, 1869. Folsom, Cook, and Peterson spent 36 days in the wilderness and observed features and phenomena not seen previously by any white man.[24]

The Folsom-Cook Expedition was also the first expedition which entered the Yellowstone to have its discoveries published, although this was not an easy task. The manuscript was rejected by *Lippincott's Magazine* with the reply that they did not print fiction.[25] *Harper's Magazine* also rejected the manuscript claiming that the statements were incredulous.[26] The manuscript was finally accepted and published by the *Western Monthly*,[27] albeit abbreviated and altered, as the editor cut out portions of Folsom's and Cook's diary.[28]

Following the expedition, David Folsom worked for the Surveyor General of Montana, Henry Washburn, and worked closely with Walter deLacy.[29] deLacy took the information Folsom and Cook gathered and revised the 1869 *Map of the Territory of Montana with Portions of the Adjoining Territories* and published the *deLacy Map of 1870.* The *deLacy Map of 1870* was the most accurate map published to date. The shape of Yellowstone Lake shows the West Thumb and an arm to the south (no distinction of the South or Southeast Arm is made). The location of the Upper and Lower falls on the Yellowstone River is clearly noted. Their heights of 115 feet and 350 feet, respectively, are also listed. These calculations were very close to today's accepted measurements of 109 feet and 308 feet.[30] Folsom and Cook's route is also clearly identified.[31] It was the *deLacy Map of 1870* that Henry Washburn took with him when he explored the Yellowstone in 1870.

David Folsom's and Charles Cook's contributions to the discovery and exploration of the Yellowstone country are remarkable considering the size of their expedition, only three men. Their accomplishments were significant: they were the first to have their findings published; their information resulted in a much improved and accurate map; David Folsom was the second person to suggest that this area be set aside and preserved; and the success of their expedition was encouragement for the Washburn-Doane Expedition in 1870.[32]

Washburn-Doane Expedition, 1870

The success of the Folsom-Cook Expedition and the publication of their experiences in the *Western Monthly* increased interest in the exploration of the Yellowstone country. Henry Washburn was especially interested in exploring the Yellowstone.

Washburn began recruiting men in the spring of 1870 and by August his party consisted of 19 men, including many Montana celebrities. The party included: Washburn, surveyor general for the Montana Territory; Nathaniel P. Langford, collector of internal revenue for Montana and later the first superintendent of Yellowstone National Park; Samuel Hauser, civil engineer and president of the First National Bank of Helena; Cornelius Hedges, judge and the third person to suggest that Yellowstone National Park be established; Walter Trumbull, assistant assessor of internal revenue for Montana; and Truman Everts, the oldest member of the party, former assessor of internal revenue for Montana, and whose claim to fame was his being lost 37 days during the expedition and surviving the experience.[33]

Washburn learned by chance that Lieutenant Gustavus Doane was very interested in his planned expedition to the Yellowstone country. Hauser and Washburn sent a telegram to General Hanncock in St. Paul requesting that Lieutenant Doane accompany the party. A reply was received on August 14, 1870, from General Hanncock with orders that Lieutenant Doane be the officer in charge of the military escort which would accompany the Washburn party.[34,35] This was the first military expedition to enter Yellowstone National Park.

Washburn and his expedition departed Helena on August 17, 1870. They joined up with Lieutenant Doane and five soldiers on August 22 at Fort Ellis and the Washburn-Doane Expedition, often referred to as the Yellowstone Expedition, was complete.[36] They crossed the present day northern boundary of Yellowstone National Park on August 26[37] and Langford's last entry in his diary was dated September 27, 1870.[38] The expedition was gone 42 days. During that time the party saw more than any previous expedition and added a considerable amount to the knowledge of the Yellowstone country. Its members published a diary, an official report, and a number of magazine articles.[39]

The Yellowstone Expedition, like Folsom and Cook, traveled up the Yellowstone River to Yellowstone Lake. At this point the party preceded south along the eastern shore of the lake and passed the South and Southeast Arms and West Thumb of the lake. They were the first to explore and map this area of Yellowstone Lake. The expedition then entered the Upper Geyser Basin, followed the Firehole River to the Madison River, exited the park on the west side, and proceeded north on their return to Helena.[40]

Lieutenant Doane not only led the first military expedition into the park but his report was also the first, official government report written about what we know as Yellowstone National Park. Doane's report is excellent; it is readable, accurate, and concise. The introductory material to volume two of Bonney's *Battle Drums and Geysers* includes the following quotes by Hiram Chittenden and Ferdinand Hayden about Doane's report:

> Doane's part in the expedition of 1870 is second to none. He made the first official report upon the wonders of Yellowstone and his fine descriptions have never been surpassed by any subsequent reports. H. M. Chittenden.
>
> Doane's journal has not been surpassed by any official report made to our government since the times of Lewis and Clark. F. V. Hayden.[41]

Henry Washburn and Lieutenant Gustavus Doane each compiled a map of the route of the expedition. However, neither map was published with their respective reports but both appear in Bonney's book.[42] Both maps show the route of the expedition and a more accurate Yellowstone Lake is presented. The inaccuracies which appeared on the *deLacy Map of 1870* appear on Washburn's and Doane's maps: Yellowstone Lake is too far south and east; Jackson Lake is too far west; two lakes are shown between Yellowstone Lake and the Firehole Basin where only one exists.[43]

The Yellowstone Expedition discovered and named many of the geysers in the Upper Geyser Basin, including Old Faithful.[44] They were the first to encircle Yellowstone Lake and, as a result, produce a more accurate map of the lake's shoreline. Their diaries, journals, reports, and articles were very detailed and vivid in the description of what they saw and where they saw it. Their overall contribution to the exploration of the Yellowstone country was significant and instrumental in the establishment of Yellowstone National Park.

The next expedition to enter the park area was the combined expedition of Ferdinand V. Hayden and Captains John Barlow and David Heap in 1871. This was the last expedition to explore the wonders of Yellowstone before the establishment of Yellowstone National Park in 1872.

Hayden and Barlow-Heap Expedition, 1871

Ferdinand V. Hayden was not unfamiliar with the Yellowstone country when he led his 1871 expedition.[45] Previously he had been part of the expedition to the Lower Yellowstone led by G. K. Warren in 1856 and the Raynolds' Expedition of 1859-1860.[46] These near successes of entrance to the Yellowstone country and the reports and maps recently published by Folsom, Cook, Washburn, Doane, and deLacy were more than enough encouragement for Hayden to achieve his goal of entering this geologic wonderland.

Hayden was in charge of the newly created United States Geological Survey of the Territories, which was later merged with the King and Powell surveys to form the United States Geological Survey in 1879.[47] The 1871 expedition was the first of three that Hayden would lead into the Yellowstone.[48] It was also the first government sponsored expedition to successfully explore the area, the first to successfully photograph the area, and the first to publish a map of the newly established Yellowstone National Park when the official report was published in 1872.[49]

Hayden's expedition was accompanied by a simultaneous expedition of the United States Corps of Engineers led by Captain John W. Barlow, assisted by Captain David P. Heap.[50] The two parties explored separately and together during the several weeks in the region. The two expeditions consisted of over 70 men. Hayden's party totaled 34 men, including: Thomas Moran, artist; William H. Jackson, photographer; Anton Schoenborn, topographer; and Albert C. Peale, mineralogist.[51] The Barlow-Heap party, a smaller expedition of 11 men, included: Thomas Hine, photographer; H. G. Prout, assistant topographer; and W. H. Wood, draughtsman.[52] Hayden and Barlow shared a military escort who was later replaced by Lieutenant Gustavus Doane, no stranger to the Yellowstone region.[53] In addition, Barlow was assigned a separate military escort of six men which was under his direct command.[54]

Both expeditions originated at Fort Ellis, Montana. Hayden departed on July 15, 1871. Barlow and Heap departed the next day. Their routes were somewhat similar to Washburn and Doane's route. However, due to the larger size and length of time of the expeditions, each party covered more territory, discovered and named more features, and returned with more specimens, photographs and astronomical observations than any previous expedition. They made a complete loop of Yellowstone Lake and explored far south of the lake (actually exiting what is the park's southern boundary). Hayden's party returned to Fort Ellis on August 30, 1871, while Barlow's party reached the fort on September 1, 1871.[55]

The expedition completed, Hayden returned to Washington, Barlow to Chicago, and Heap to St. Paul. The men began to compile the data and write their reports. However, misfortune struck. Hayden's topographer, Anton Schoenborn, committed suicide on the trip back to Washington, and United States Coast Survey personnel compiled the maps for Hayden's report from Schoenborn's field notes. Barlow had been in Chicago only a few days when the great fire of October 8-11, 1871, destroyed almost all of his material from the expedition:

notes, biological specimens, meteorological records, and all but 16 of Hine's photographic plates. Fortunately, Heap's astronomical records were in St. Paul.[56]

Despite these losses, Hayden's and Barlow's reports and maps were thorough, detailed, and accurate. Their maps were exquisite. Hayden's map, *Yellowstone National Park, From Surveys Made under the Direction of F. V. Hayden,* accompanied his report (fig. 20)[57] and was later published in the article he wrote for the *American Journal Science and Arts.*[58] This is the first map that shows the boundaries of Yellowstone National Park. The shape of Yellowstone Lake is very accurate. The location, size and shape of Madison Lake (present day Shoshone Lake) and Heart Lake are relatively accurate. However, present day Lewis Lake, located between Madison and Heart Lakes, is omitted and the non-existent De Lacy's Lake still appears on Hayden's map at the southwest corner of map, just outside the park's boundaries.

Barlow's map, *Sketch of the Yellowstone Lake and the Valley of the Upper Yellowstone River,*[59] was the first map drawn of the area through the use of triangulation. This made it the first map to show features and details in their proper relationship to longitude and latitude.[60] The Barlow-Heap map shows an even more accurately drawn Yellowstone Lake. Madison Lake (present day Shoshone Lake) is properly positioned and shows it draining into an unnamed lake (present day Lewis Lake). Heart Lake appears in its proper location but an unnamed, non-existent lake appears between Heart and the unnamed lake (Lewis Lake). Barlow's non-existent lake is similar in size and shape to Hayden's De Lacy's Lake but is in a completely different location. This error is a result of previous expeditions and inaccurate observations. Hayden corrected this on an 1876 map of the park published in *The Yellowstone National Park, and the Mountain Regions of Idaho, Nevada, Colorado and Utah*[61] and on the 1878 *Map of Yellowstone National Park, Showing Distribution of Hot Springs.*

The results of these two expeditions are extensive and unprecedented. Their reports and related publications provided the nation with a comprehensive written narrative of the Yellowstone wonderland.[62] Hayden's and Barlow's maps were the most accurate and detailed to date. Jackson's photographs and Moran's drawings pro-

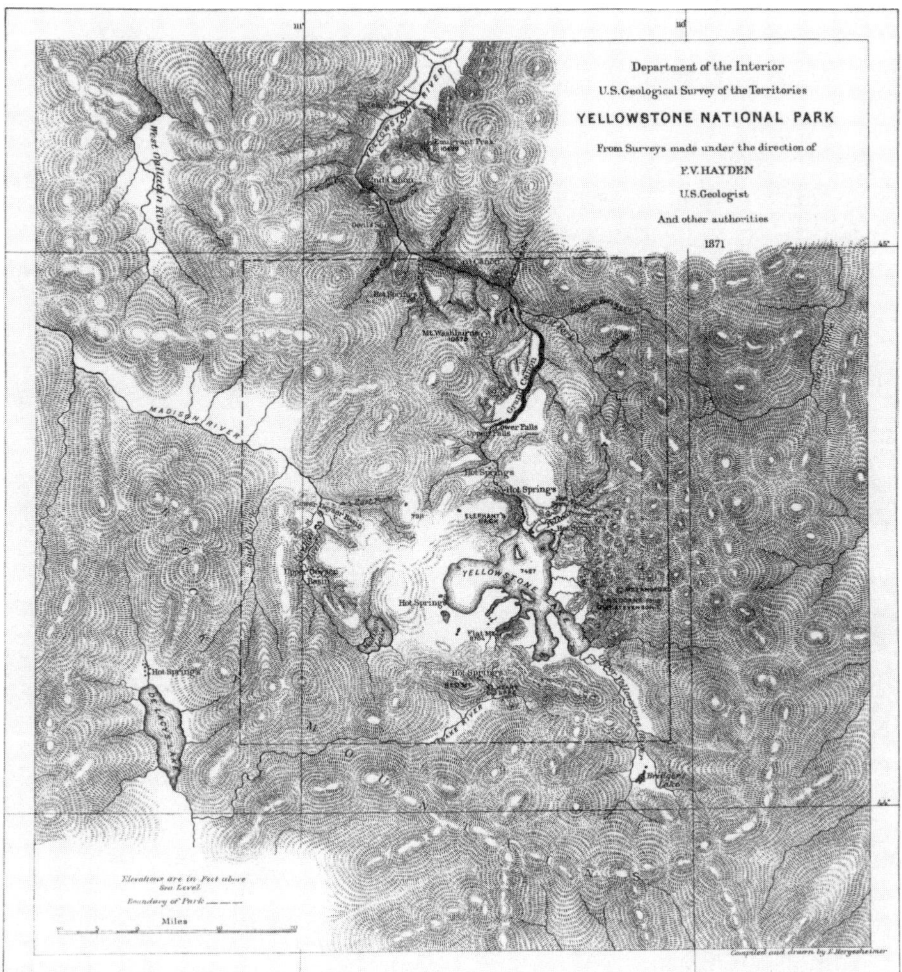

Fig. 20. From Hayden's *Preliminary Report,* 1872. (Courtesy of Tufts University)

vided a visual picture of the region for the first time. All of these factors, coupled with previous expeditions, reports and maps, were the final motivation leading to the Yellowstone Act of 1871 and the establishment of Yellowstone National Park, the world's first national park.

THE ESTABLISHMENT OF YELLOWSTONE NATIONAL PARK

The literature published on the establishment of Yellowstone National Park is extensive.[63] Although Yellowstone National Park was the nation's first national park, the idea of setting aside land as a national preserve was not new. The Hot Springs Reservation in Arkansas was established in 1832. In 1864 the Federal government ceded land to the state of California and the Yosemite Valley and Mariposa Big Tree Grove was established.[64] Hot Springs and Yosemite were later established as national parks in 1921 and 1890, respectively.[65]

Senator Clarke Pomeroy of Kansas introduced Senate Bill 392 on December 18, 1871, and William Clagett of Montana introduced House Bill 764 on the same date. It was the Senate bill, with revisions, that President Grant signed on March 1, 1872, and the establishment of Yellowstone National Park was completed.[66]

It is safe to say that no one individual can be credited with the discovery of the region or the passage of the act. The establishment of Yellowstone National Park was truly a joint effort, albeit unintentional. It was unfortunate that many individuals, Langford and Hayden included, claimed to be the discoverer of the park and the main reason for its establishment.[67] Each and every man associated with the exploration, mapping, and legislation of Yellowstone National Park, from John Colter to President Grant, is responsible for the establishment of the first national park.

Nathaniel Langford was appointed as the park's first superintendent on May 10, 1872, a position he held for nearly five years. It was a time when the park received no appropriated monies and was not developed nor protected.[68] During the early years of Yellowstone

National Park it was "a 'public park or pleasuring ground' only in the wording of the statute."[69]

Exploration continued after the park was established, and the next eight years saw five more expeditions enter the park and add to the existing literature and refine and correct earlier maps.

POST-PARK EXPEDITIONS, 1872-1879

Hayden Expedition, 1872

In 1872 Ferdinand V. Hayden completed his second exploration of Yellowstone National Park and his sixth as director of the United States Geological Survey of the Territories.[70] This was largest expedition to date. Hayden divided his expedition into Northern and Southern divisions which resulted in covering more ground than ever before. The Southern Division explored the southern shores of Shoshone Lake and the northern and eastern shores of yet to be named Leigh Lake.[71]

Hayden's report contained five maps, none of which was a map of the entire park,[72] included detailed maps of the Shoshone Geysers and the sources of the Snake River.[73] Hayden did not explore Yellowstone National Park again until 1878. Between 1872 and 1878 two major military expeditions, each led by different individuals, explored the park.

Jones Expedition, 1873

Captain William A. Jones led a Corps of Engineers expedition to Yellowstone National Park which was part of a larger exploration of northwestern Wyoming for military defenses.[74] Included in Jones' party were Charles Parry, botanist, and Theodore Comstock, geologist. Comstock's two part article on the scientific value of the park and its improvement set standards for future exploration and scientific reporting.[75]

Jones' approach to the park was different than previous expeditions. Jones was the first to successfully enter the park from the east when he discovered and named Jones Pass. He also left the park through the south and discovered Togwotee Pass which crossed the Continental Divide.[76] This was the pass Raynolds was unable to find in his 1859-1860 expedition. One other discovery of significance was the verification of Two Ocean Water. This is where the water drains into the Pacific and Atlantic water sheds simultaneously. Jim Bridger had insisted for years that such a place existed.[77]

The maps which accompanied his report were numerous. A map of Western Wyoming was published; 49 detailed trail-maps were published; and one of the first colored, geological maps of a portion of Wyoming was also published.[78]

Ludlow Reconnaissance, 1875

Captain William Ludlow began his Yellowstone reconnaissance on June 30, 1875, when he left the Corps of Engineers headquarters in St. Paul, Minnesota. It was 45 days later on August 14, 1875, when Ludlow and his party of 12 men entered Yellowstone National Park.[79] The Ludlow reconnaissance was small and spent only 13 days in the park[80] but his findings and observations were significant.

He returned with the most accurate measurement of the Upper and Lower falls of the Yellowstone, although the Upper Falls was measured at 110 feet, one foot too high, and the Lower Falls at 310 feet, two feet too high.[81] His report included a map of the entire reconnaissance area, a map of Judith Basin in Montana, and a very detailed map of the Upper Geyser Basin in Yellowstone National Park. This last map was drawn to a scale of 1:6,000. It shows the course of the Firehole River and accurately locates and names the prominent geysers in the basin (Old Faithful, Giantess, Bee Hive, Castle, Grand, Giant, Grotto, Pyramid and Punch Bowl).[82]

What was most significant about Ludlow's expedition and report was his appreciation of the fragile environment of Yellowstone National Park. He realized that it was slowly being destroyed. Ludlow outlined a plan of conservation and protection in his report which would save the park from future destruction.[83] His proposals were eventually instituted in the mid-1880s when the Corps of

Engineers began road construction and the Cavalry patrolled the park.[84]

Hayden Expedition, 1878

Hayden led his last expedition to Yellowstone National Park in 1878. It was also the twelfth and last exploration under the auspices of Hayden's United States Geological and Geographical Survey of the Territories.[85] The following year would see the consolidation of the Hayden, Wheeler, King, and Powell surveys and the creation of the United States Geological Survey.[86] This twelfth report was published in 1883, four years after the survey was established.

This expedition included many of the men from the 1871 expedition: William Henry Jackson, Albert C. Peale, and James Stevenson.[87] Unlike the 1871 expedition, which was primarily an exploration of the Yellowstone country, the 1878 expedition explored large areas in Wyoming and Idaho, in addition to Yellowstone National Park. It was by far the most voluminous of all of Hayden's reports: two printed volumes, nearly 1400 pages in length, over 300 plates, and 23 maps, 10 of which were published in a separate case.[88] Eleven of the maps pertain to Yellowstone National Park and its features, primarily the hot springs and geyser basins. Individual maps of the springs at Crater Hills, Mud Volcanoes and Witch Creek, Mammoth Hot Springs, Egeria Springs, Gibbon Geyser Basin, Lower Geyser Basin, Upper Geyser Basin, Shoshone Geyser Basin and the Rustic Geyser Group are included.[89]

The last map in Hayden's report to be discussed is the *Map of Yellowstone National Park, Showing Distribution of Hot Springs* (fig. 21).[90] Many of the inaccuracies of earlier maps, including Hayden's 1871 map, have been corrected. De Lacy's Lake, the non-existent lake on the 1871 map, has been deleted. Madison Lake is now named Shoshone Lake, and Leigh and Heart lakes are in their proper locations and their shape and size are correct. Improvements on the shape and size of Yellowstone Lake are also apparent. More named features and accurate elevations also appear on the 1878 map.

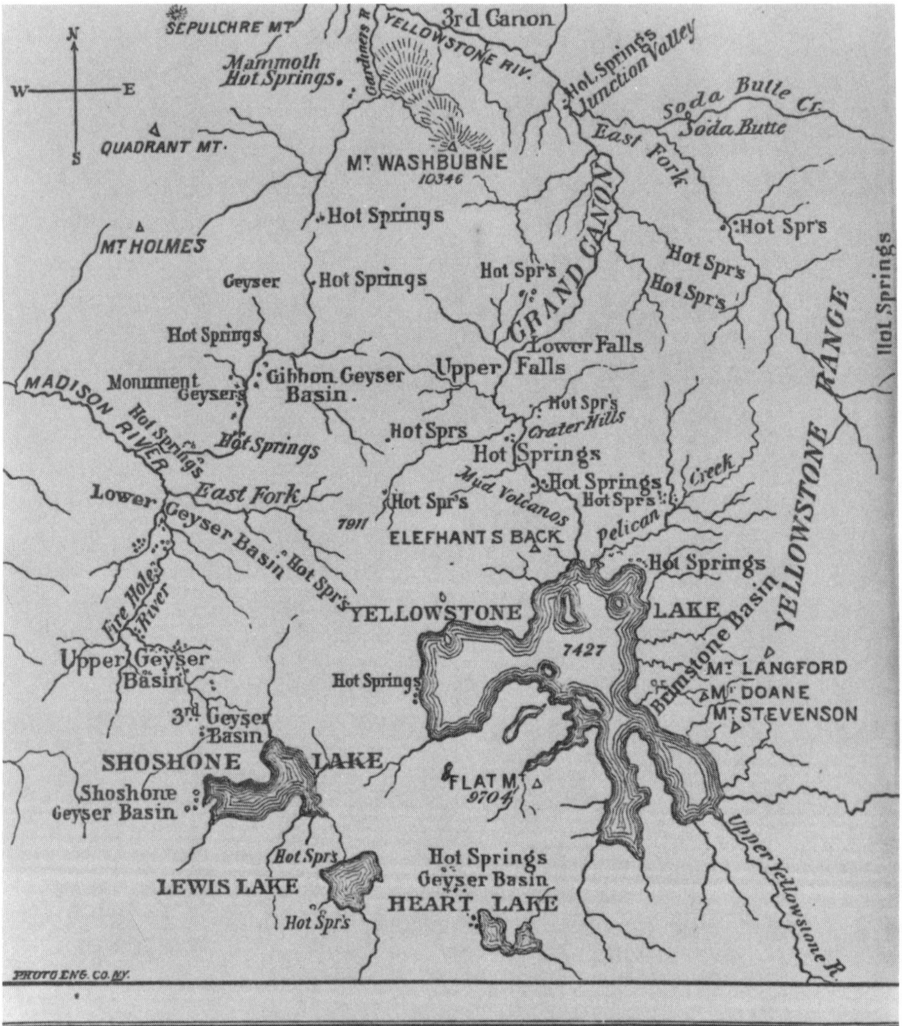

Map of Yellowstone National Park, showing distribution of Hot Springs.

Fig. 21. The map of 1878 from Hayden's 1883 report. (Courtesy of Tufts University)

This expedition marked an end to an era of exploration in Yellowstone National Park. Lieutenant General Philip Sheridan led an expedition in 1881 to the park but his report added nothing significant to the existing record.[91] The major exploration of Yellowstone National Park was completed. The park was established. People throughout the country were aware of the beauty and natural phenomena that existed in the park. This was a direct result of the official reports, magazine articles, photographs, and maps which had been published since the Lewis and Clark Expedition.

Now, the mapping, field work, and continued exploration and study of the United States, including Yellowstone National Park, would be under the control of the newly created United States Geological Survey.

UNITED STATES GEOLOGICAL SURVEY, 1879-1989

On March 3, 1879, President Rutherford B. Hayes signed House Bill 6471 and the United States Geological Survey was established. Clarence King, who had been in charge of the War Department's United States Geological Exploration of the Fortieth Parallel, was confirmed as the survey's first director on April 3, 1879, by the Senate.[92]

This marked a new era in mapping the country with an organized and systematic mapping program. This would have a positive effect on the mapping of Yellowstone National Park, an effect seen in the first topographic map of the park published by the Geological Survey.

The *Topographical Map of Yellowstone National Park and North Western Wyoming* was published at a scale of 1:125,000. The publication date of this map is questionable, either 1885 or 1886.[93] It was printed on a single sheet and was the largest scale, single sheet map of the entire park ever published. It included more detail and place names and provided the user with topography of the entire area. This particular map has been revised many times since 1885-1886, the latest revision being 1961. A shaded relief edition of the topographic is also available; it, too, was last revised in 1961.

The next major map published by the Geological Survey was the *Yellowstone National Park, Wyoming* in 1896.[94] This was Folio 30 of the *Geologic Atlas of the United States.* The folio consisted of six pages of text, three leaves of photographs, and eight maps (four topographic and four geologic). The colored geologic maps used the topographic maps as a base.

The sheet names, beginning with the northwest sheet and going clockwise, are: *Gallatin, Canyon, Lake,* and *Shoshone.* These four sheets, along with the *Crandall* and *Ishawooa* sheets found in folio 52,[95] provided complete coverage the entire park at a scale of 1:125,000. The four sheets from Folio 30, in essence, were the 1885-1886 *Topographical Map of Yellowstone National Park and North Western Wyoming,* with minor revisions, cut into four sectional maps. The margins of the Yellowstone folio maps indicate that the area was surveyed in 1883-1885 and that these maps are the "Edition of April 1896." These maps were the only such maps available until the 15-minute series maps of Yellowstone National Park were published between 1940 and 1959.

The Geological Survey published Arnold Hague's *Geology of the Yellowstone National Park* and accompanying atlas in 1904.[96] The atlas was folio size and contained 27 sheets, of which, four topographic and four geologic maps were from the 1896 *Yellowstone National Park, Wyoming.* However, these eight sheets had a 1904 publication date.

Hague's monograph included many large scale geological maps. Some of these were: Mammoth Hot Springs (1:14,000); Norris Geyser Basin (ca 1:3,600); Firehole Geyser Basin (ca 1:9,250); Excelsior Geyser Basin (ca 1:2,400); Upper Geyser Basin (1:6,000); Shoshone Geyser Basin (ca 1:1,700); and the shores of Yellowstone Lake (ca 1:85,000).

The last Geological Survey mapping of Yellowstone National Park to be discussed are the 1:24,000-scale map series (7.5-minute maps); 1:62,500-scale map series (15-minute maps); 1:100,000-scale map series (metric maps); and 1:250,000-scale map series (1 x 2 degree maps). The 1:250,000-scale map series, previously published by the Army Map Service, are now published and revised by the Geological Survey. Four sheets provide complete coverage of Yellowstone National Park at this scale. The sheet names and the date of the last revision are: *Bozeman* (1972); *Billings* (1979); *Cody* (1980); and *Ashton* (1972).

The 1:100,000-scale map series is the newest of the four series and is the first and only series to provide coverage of Yellowstone National Park at metric scale. Two sheets, *Yellowstone National Park North* (1983) and *Yellowstone National Park South* (1982), provide complete coverage.

The 30-minute topographic maps from the *Yellowstone National Park, Wyoming* (Folio 30) and *Absaroka, Wyoming* (Folio 52) were the only sheet maps available for Yellowstone National Park until 1940 when the first of 26, 15-minute topographic maps were published. It would take 20 years to complete the mapping and coverage of Yellowstone National Park at this 1:62,500 scale.

It was not until 1986 when the first 7.5-minute topographic map was published and Yellowstone National Park finally had mapping coverage at the scale of 1:24,000. The mapping at this scale is now complete for the park; all 77 topographic maps are Geological Survey provisional edition maps. It took approximately 180 years...from when John Colter first entered the Yellowstone country in 1807 to 1986 when the publication of the first large scale maps of Yellowstone National Park was completed...to achieve the detail, accuracy, and continuity that each and every explorer had as his goal.

Over the years the Geological Survey has published *Professional Papers, Circulars, Bulletins, Water-Supply Papers,* miscellaneous publications, and maps and atlases which studied and mapped Yellowstone National Park and specific areas within the park. There are as many, if not more, non-Geological Survey publications and maps on Yellowstone National Park. Aerial photography, infrared and satellite imagery and digitized mapping data also exist for Yellowstone National Park.

190 years of exploration and mapping of Yellowstone National Park have produced the events, publications, and maps that have had an influential role and played an integral part in the establishment and ongoing exploration of Yellowstone National Park. One final map needs to be mentioned, the *Yellowstone and Grand Teton* map published by the National Geographic Society.[97] Of the highest quality, the roads and trails shown on this map are very similar to routes of Folsom, Cook, Washburn, Doane, Hayden, Barlow, Jones, and

Ludlow. Their legacy lives on. This is the first in a new National Geographic Society map series called "Special Places of the World Map Series." It is appropriate and fitting that the first "special place" to be mapped in this series is Yellowstone National Park...the world's first national park!

NOTES

*This paper is based on a presentation given at the American Library Association Annual Conference, New Orleans, Louisiana, June 1988, and is a revision of the article that was published in *Meridian* No. 3 (1990): 5-21.

1. William F. Raynolds, *Report of Brevet Brigadier General W. F. Raynolds on the Exploration of the Yellowstone and the Country Drained by that River*, 40th Cong., 1st sess., 1868, S. Ex. Doc. 77. (Serial 1317), 11. [Also issued as a publication of the Engineer Department, War Department, W 7.5: R21.]
2. Gustavus C. Doane, *Report of Lieutenant Gustavus C. Doane upon the So-Called Yellowstone Expedition of 1870*, 41st Cong., 3rd sess., 1871, S. Ex. Doc 51. (Serial 1440), 37.
3. Ferdinand V. Hayden, "The Wonders of the West...II: More about the Yellowstone," *Scribner's Monthly* 3 (February 1872): 396.
4. "An Act to Set Apart a Certain Tract of Land Lying Near the Headwaters of the Yellowstone River as a Public Park," *Statutes at Large and Proclamations of the United States of America* (Boston: Little, Brown, and Co., 1873), 17: 32-33.
5. Aubrey L. Haines, *The Yellowstone Story: A History of Our First National Park* (Yellowstone, Wyoming: Yellowstone Library and Museum Association, 1977), 1: 103.
6. Numerous books have been written on the exploration and history of Yellowstone National Park, each providing a detailed and accurate account. Individuals wishing more information about Yellowstone National Park should read: Kenneth Baldwin, *Enchanted Enclosure: The Army Engineers and Yellowstone National Park* (Washington: U.S. Government Printing Office,

1976); Hiram M. Chittenden, *The Yellowstone National Park* (Cincinnati: Robert Clarke Co., 1905), or previous or subsequent editions; *Ferdinand Vandiveer Hayden and the Founding of Yellowstone National Park* (Washington: U.S. Government Printing Office, 1973); Aubrey L. Haines, *Yellowstone National Park: Its Exploration and Establishment* (Washington: U.S. Government Printing Office, 1974); Aubrey L. Haines, *The Yellowstone Story: A History of Our First National Park*, 2 vols. (Yellowstone National Park, Wyoming: Yellowstone Library and Museum Association, 1977); David A. Clary, *"The Place Where Hell Bubbled Up:" A History of the First National Park* (Washington: U.S. Government Printing Office, 1972); John H. Raftery, *Historical and Descriptive Sketch of the Yellowstone National Park*, 60th Cong., 2nd sess., 1909 S. Doc. 752 (Serial 5409); Louis C. Cramton, *Early History of Yellowstone National Park and Its Relation to National Park Policies* (Washington: U.S. Government Printing Office, 1932); or William T. Jackson, *The Early Exploration and Founding of Yellowstone National Park*, Ph.D. diss. (Austin: University of Texas, 1940). Many of the maps discussed in this paper can be found in the accompanying reports or reprints which are cited in this paper. Some of the maps are also reprinted in Carl I. Wheat, *Mapping the Transmississippi West, 1540-1861*, 5 vols. in 6 vols. (San Francisco: Institute of Historical Cartography, 1957-1963).

7. Herman I. Viola, *Exploring the West* (Washington: Smithsonian Books, 1987), 18-20; *Exploring the American West, 1803-1879*, National Parks Handbook No. 116 (Washington: U.S. Government Printing Office, 1982), 22-26.

8. Chittenden, *The Yellowstone National Park*, 15-16; Clary, *"The Place Where Hell Bubbled Up,"* 9; Cramton, *Early History of Yellowstone National Park*, 5; Haines, *Yellowstone National Park*, 3-4.

9. Chittenden, *The Yellowstone National Park*, 26; Jackson, *The Early Exploration and Founding of Yellowstone*, 27-41; Haines, *Yellowstone National Park*, 4-5; Haines, *The Yellowstone Story*, 1: 5.

10. Haines, *Yellowstone National Park*, 185-186; Haines, *The Yellowstone Story*, 1: 34.

11. William H. Goetzmann, *Army Exploration in the American West, 1803-1863* (New Haven: Yale University Press, 1959) 25, 106; Haines, *Yellowstone National Park*, 17, 21-23, 187-189; Viola, *Exploring the West*, 140-141; *Exploring the American West*, 70-72.

12. Haines, *The Yellowstone Story*, 1: 86.

13. Baldwin, *Enchanted Enclosure*, 5.

14. Haines, *The Yellowstone Story*, 1: 87; Baldwin, *Enchanted Enclosure*, 5.

15. Raynolds, *Report of Brevet Brigadier General W. F. Raynolds*, 10-11.

16. Goetzmann, *Army Exploration in the American West*, 411.

17. Chittenden, *The Yellowstone National Park*, 60.

18. Haines, *Yellowstone National Park*, 26.

19. Haines, *The Yellowstone Story*, 1: 89.

20. Haines, *Yellowstone National Park*, 136-137.

21. Ibid., 55-56, 139, 191-194; Haines, *The Yellowstone Story*, 1: 65.

22. Haines, *Yellowstone National Park*, 55-56, 194.

23. Raftery, *Historical and Descriptive Sketch*, 7; Charles W. Cook, "The Valley of the Upper Yellowstone," *Western Monthly* 4 (July 1870): 60-61.

24. Ibid., 61-67; Jackson, *The Early Exploration and Founding of Yellowstone*, 127-137; Haines, *The Yellowstone Story*, 1: 93-103.

25. Baldwin, *Enchanted Enclosure*, 1.

26. Raftery, *Historical and Descriptive Sketch*, 8.

27. Cook, "The Valley of the Upper Yellowstone," 60-67.

28. Haines, *Yellowstone National Park*, 45-46.

29. Jackson, *The Early Exploration and Founding of Yellowstone*, 141-142; Haines, *The Yellowstone Story*, 1: 101-102.

30. Ibid., 98, 341.

31. Haines, *Yellowstone National Park*, 194.

32. Ibid., 56.

33. Nathaniel P. Langford, *Diary of the Washburn Expedition to the Yellowstone and Firehole Rivers in the Year 1870* (St. Paul: Langford, 1905), xvi; Jackson, *The Early Exploration and Founding of Yellowstone*, 146-159; *Yellowstone Wonderland: The Exciting Beginning of Yellowstone National Park*, Heritage Series No. 8 (Helena: Montana Historical Society, 1960), 3-4.

34. Jackson, *The Early Exploration and Founding of Yellowstone*, 166-170; *Yellowstone Wonderland*, 5-6.

35. *Yellowstone Wonderland*, 5-6.

36. Langford, *Diary of the Washburn Expedition*, 1, 4.

37. *Yellowstone Wonderland*, 8.

38. Langford, *Diary of the Washburn Expedition*, 120.

39. The published reports, journals, and magazine articles include: Nathaniel P. Langford, "The Wonders of the Yellowstone," *Scribner's Monthly* 2 (May-June 1871): 1-17, 113-128; Nathaniel P. Langford, "The Ascent of Mount Hayden," *Scribner's Monthly* 6 (July 1873): 129-157; Nathaniel P. Langford, *Diary of the Washburn Expedition to the Yellowstone and Firehole Rivers in the Year 1870* (St. Paul: Langford, 1905); Nathaniel P. Langford, *The Discovery of Yellowstone National Park, 1870*, 2nd edition (St. Paul: J. E. Haynes, 1923); Nathaniel P. Langford, *The Discovery of Yellowstone National Park: Journal of the Washburn Expedition to the Yellowstone and Firehole Rivers in the Year 1870* (Lincoln: University of Nebraska Press, 1972); Walter Trumbull, "The Washburn Yellowstone Expedition," *Overland Monthly* 6 (May-June 1871): 431-437, 489-496; Truman C. Everts, "Thirty-Seven Days of Peril," *Scribner's Monthly* 3 (November 1871): 1-17; Cornelius Hedges, "Journal of Judge Cornelius Hedges," *Contributions to the Historical Society of Montana* 5 (1904): 370-394; and Gustavus C. Doane, *Report of Lieutenant Gustavus C. Doane upon the So-Called Yellowstone Expedition of 1870*, 41st Cong., 3rd sess., 1871, S. Ex. Doc. 51. (Serial 1440).

40. Orin H. and Lorraine Bonney, *Battle Drums and Geysers* (Houston: Bonney and Bonney, 1970), 2: 199, *Map of the Route of the Yellowstone Expedition . . . September 1870.*

41. Ibid., 2: unnumbered page before title page.

42. Ibid., 2: 199, *Map of the Route of the Yellowstone Expedition . . . September 1870*, 2: 295, *Route of the Washburn Party 1870.*

43. Ibid., 2: 399-403; Haines, *Yellowstone National Park*, 92-93, 195-196.

44. Langford, *Diary of the Washburn Expedition*, 106-112; Doane, *Report of Lieutenant Gustavus C. Doane*, 28-32.

45. Haines, *The Yellowstone Story*, 2: 438.

46. Ferdinand V. Hayden, *Preliminary Report of the United States Geological Survey of Montana and Portions of Adjacent Territories; Being a Fifth Report of Progress*, 42nd Cong., 2nd sess., 1872 H. Ex. Doc. 326 (Serial 1520), 7. [Also issued as a publication of the Geological and Geographical Survey of the Territories, I 18.1: 871.]; *Ferdinand Vandiveer Hayden and the Founding of Yellowstone National Park* (Washington: U.S. Government Printing Office, 1973), 5.

47. Haines, *The Yellowstone Story*, 2: 438.

48. Haines, *Yellowstone National Park*, 143.

49. Ibid., 100; Haines, *The Yellowstone Story*, 1: 141; Ferdinand V. Hayden, "The Yellowstone National Park," *American Journal of Science and Arts* Third Series 3 (January-June 1872): 294.

50. Baldwin, *Enchanted Enclosure*, 15.

51. *Ferdinand Vandiveer Hayden*, 8; Haines, *The Yellowstone Story*, 1: 141-142.

52. John W. Barlow and David P. Heap, *Report of Reconnaissance of Basin of Upper Yellowstone in 1871*, 42nd Cong., 2nd sess., 1872 S. Ex. Doc. 66 (Serial 1479), 3, 5. [Also issued as a publication of the Engineer Department, War Department, W 7.5: B 24.]

53. Hayden, *Preliminary Report . . . Being the Fifth Annual Report of Progress*, 6.

54. Haines, *The Yellowstone Story*, 1: 144.

55. Ibid., 1: 145, 190; Excellent narratives on the Hayden and Barlow-Heap Expedition can be found in: Baldwin, *Enchanted Enclosure*, 14-23; Richard A. Bartlett, *Great Surveys of the American West* (Norman: University of Oklahoma Press, 1962), 35-73; *Ferdinand Vandiveer Hayden*; Haines, *The Yellowstone Story*, 1: 142-151; and Haines, *Yellowstone National Park*, 99-106 in addition to the official reports of Hayden, *Preliminary Report . . . Being a Fifth Annual Report of Progress,* and Barlow, *Report of Reconnaissance.*

56. Haines, *The Yellowstone Story*, 1: 153.

57. Hayden, *Preliminary Report . . . Being a Fifth Annual Report of Progress*, 162.

58. Hayden, "The Yellowstone National Park," 294-297.

59. Barlow, *Report of Reconnaissance*, located at the end of the report; Baldwin, *Enchanted Enclosure*, 17.

60. Haines, *The Yellowstone Story*, 1: 153.

61. Wheat, *Mapping the Transmississippi West* 2: 351-352, 433, Wheat map number 1269, *Map of the Yellowstone National Park showing the location of views.*

62. The reports and narratives of Hayden's and Barlow and Heap's expeditions can be found in: Ferdinand V. Hayden, "The Hot Springs and Geysers of the Yellowstone and Firehole Rivers," *American Journal of Science and Arts* Third Series 3 (January-June 1872): 105-115; Hayden, "The Wonders of the West;" Hayden, "The Yellowstone National Park;" and Barlow, *Report of Reconnaissance.*

63. For detailed coverage on the creation and legislative process of the Yellowstone Act of 1872 please consult: Chittenden, *The Yellowstone National Park*; Clary, *"The Place Where Hell Bubbled Up;"* Cramton, *Early History of Yellowstone National Park*; Haines, *The Yellowstone Story; Yellowstone Wonderland* (Montana Historical Society); and Raftery, *Historical and Descriptive Sketch.*

64. *The National Parks: Shaping the System,* Part 2 (Washington: U.S. Government Printing Office, 1985), 10.

65. *The National Parks: Index 1985* (Washington: U.S. Government Printing Office, 1985), 24, 28.

66. Cramton, *Early History of Yellowstone National Park,* 24-28.

67. Ibid., 28-35.

68. Haines, *The Yellowstone Story,* 2: 449.

69. Baldwin, *Enchanted Enclosure,* 85.

70. Ferdinand V. Hayden, *Sixth Annual Report of the United States Geological Survey of the Territories,* 42nd Cong., 3rd sess., 1873, H. Misc. Doc. 112 (Serial 1573). [Also issued as a publication of the Geological and Geographical Survey of the Territories, I 18.1: 872.]

71. Haines, *The Yellowstone Story,* 2: 440.

72. Hayden, *Sixth Annual Report,* xi.

73. Ibid., 244, 255.

74. Baldwin, *Enchanted Enclosure,* 45.

75. Theodore B. Comstock, "The Yellowstone National Park," *American Naturalist* 8 (February 1874): 65-79; 8 (March 1874): 155-166.

76. Haines, *The Yellowstone Story,* 1: 201; Baldwin, *Enchanted Enclosure,* 45.

77. Ibid., 45-46.

78. William A. Jones, *Report Upon the Reconnaissance of Northwestern Wyoming, Including Yellowstone National Park, Made in the Summer of 1873,* 43rd Cong., 1st sess., 1874. H. Ex. Doc. 285 (Serial 1615), v. [also issued as a publication of the Engineer Department, War Department, W 7.5: J 72.]

79. William Ludlow, *Exploring Nature's Sanctuary: Captain William Ludlow's Report of a Reconnaissance from Carroll, Montana Territory, on the Upper Missouri to the Yellowstone National Park, and Return Made in the Summer of 1875,* Engineer Historical Studies

No. 3 (Washington: U.S. Government Printing Office, 1985), vii, ix.

80. Haines, *The Yellowstone Story,* 1: 203.

81. Ibid., 357n; Ludlow, *Exploring Nature's Sanctuary,* ix, 21-22; William Ludlow, *Report of a Reconnaissance from Carroll, Montana Territory, on the Upper Missouri to the Yellowstone National Park, and Return Made in the Summer of 1875* (Washington: U.S. Government Printing Office, 1876), 21-22.

82. Ibid., 26; Ludlow, *Exploring Nature's Sanctuary,* 26.

83. Baldwin, *Enchanted Enclosure,* 67-68; Ludlow, *Report of a Reconnaissance,* 35-37; Ludlow, *Exploring Nature's Sanctuary,* 35-37.

84. Baldwin, *Enchanted Enclosure,* 68.

85. Adelaid R. Hasse, *Reports of Explorations Printed in Documents of the United States Government* (New York: Burt Franklin, 1969), 81. [Originally published as: Adelaide R. Hasse, *Reports of Explorations Printed in the Documents of the United States Government: A Contribution toward a Bibliography* (Washington: Government Printing Office, 1899) (GP 3.5: 2).]; L. F. Schmeckebier, *Catalogue and Index of the Publications of the Hayden, King, Powell, and Wheeler Surveys* (Washington: U.S. Government Printing Office, 1903), 19-20.

86. Two excellent books have been written on the creation and history of the U.S. Geological Survey. They are: Mary C. Rabbitt, *Minerals, Lands, and Geology for the Common Defence and General Welfare,* 3 vols. (Washington: U.S. Government Printing Office, 1979) and Mary C. Rabbitt, *The United States Geological Survey, 1879-1989,* U.S. Geological Survey Circular 1050 (Washington: U.S. Government Printing Office, 1989).

87. Haines, *The Yellowstone Story,* 2: 440-441.

88. Hasse, *Reports of Explorations,* 81; Schmeckebier, *Catalogue and Index,* 19-20; Ferdinand V. Hayden, *Twelfth Annual Report of the United States Geological and Geographical Survey of the Territories,* 2 vols., 47th Cong., 1st sess., 1883, H. Misc. Doc. 62 (Serial 2056-2057). [Also issued as a publication of the Geological and Geographical Survey of the Territories, I 18.1: 883/1-2.]

89. Ibid., 2: 84, 94, 106, 132, 178, 186, 240, 278, 290, 300.
90. Ibid., 2: 66.
91. Cramton, *Early History of Yellowstone National Park*, 36-37; Hasse, *Reports of Explorations*, 72.
92. Rabbitt, *Minerals, Lands, and Geology*, 1: 287-288; Richard A. Bartlett, *Great Surveys of the American West* (Norman: University of Oklahoma, 1962), xiv.
93. *The Publications of the Geological Survey, 1879-1961* lists a publication date of 1885 for this map. However, in personal correspondence (March 15, 1990), Peter Stark, who has compiled an exhaustive study and listing of U.S. Geological Survey special maps (Western Association of Map Libraries Occasional Paper No. 12), correctly points out that no publication date was listed on any of the copies that he examined while compiling his list. Only the survey dates of 1883-1884-1885 are provided. Based on this, one could assume that the map was not published until 1886. This assumption is supported further by the publication of the Geologic Atlas of the United States, Folio 30, which has the publication date of "Edition of April 1896." The four sheets that comprise Folio 30 cover the same area as the single-sheet Yellowstone map. Folio 30 is discussed and described in the next two paragraphs.
94. Arnold Hague, *Yellowstone National Park, Wyoming*, Geologic Atlas of the United States, Folio 30 (Washington: U.S. Geological Survey, 1896).
95. Arnold Hague, *Absaroka, Wyoming*, Geologic Atlas of the United States, Folio 52 (Washington: U.S. Geological Survey, 1899).
96. Arnold Hague, *Geology of the Yellowstone National Park*, U.S. Geological Survey Monograph 32 (Washington: U.S. Geological Survey, 1904); Arnold Hague, *Atlas to Accompany Monograph XXXII on the Geology of the Yellowstone National Park* (Washington: U.S. Geological Survey, 1904).
97. This map is printed as a map supplement to *National Geographic* 175 (February 1989). This issue includes the article "Yellowstone: The Fires of 1988," pages 255-273.

SELECTED CARTOBIBLIOGRAPHY

deLacy Map of 1870 [map]. Scale not given. 1870. Reduced and reprinted in: Haines, Aubrey L. *Yellowstone National Park: Its Exploration and Establishment.* Washington, D.C.: U.S. Government Printing Office, 1974, p. 194.

General Land Office Manuscript Map (Washburn's Map) [map]. Scale not given. 1869. Reduced and reprinted in: Haines, Aubrey L. *Yellowstone National Park: Its Exploration and Establishment.* Washington, D.C.: U.S. Government Printing Office, 1974, p. 193.

Lower Geyser Basin, Fire Hole River, Wyoming Territory [map]. ca 1:18,300. In: Hayden, F. V. "The Hot Springs and Geysers of the Yellowstone and Firehole Rivers," *American Journal of Science and Arts,* 3 (January-June 1872): after page 176; and U.S. House. 42nd Congress, 2nd Session. *Preliminary Report of the United States Geological Survey of Montana and Portions of Adjacent Territories* (H. Ex. Doc. 326). Washington, D.C.: U.S. Government Printing Office, 1872, p. 104. (Serial Set 1520).

Map of a Military Reconnaissance: North Western Wyoming [map]. Scale not given. 1873. Reduced and reprinted in: Baldwin, Kenneth. *Enchanted Enclosure: The Army Engineers and Yellowstone National Park.* Washington: U.S. Government Printing Office, 1976, p. 49.

Map of a Reconnaissance from Carroll Montana Ter. to the Yellowstone National Park and Return [map]. 1" = 12 mi. In: Ludlow, William. *Report of a Reconnaissance from Carroll, Montana Territory, on the Upper Missouri to the Yellowstone National Park, and Return Made in the Summer of 1875.* Washington: U.S. Government Printing Office, 1876, folded map in pocket. Reprinted in: Ludlow, William. *Exploring Nature's Sanctuary: Captain William Ludlow's Report of a Reconnaissance from Carroll, Montana Territory, on the Upper Missouri to the Yellowstone National Park, and Return Made in the Summer of 1875.* Engineer Historical Studies No. 3. Washington: U.S. Government Printing Office, 1985, folded map in pocket.

Map of the Route of the Yellowstone Expedition, 1870 [map]. 1" = 6 mi. Reprinted in: Bonney, Orrin H. and Lorraine G. Bonney. *Battle*

Drum and Geysers. 3 vol. Houston: Bonney and Bonney, 1970, vol. 2, folded map, p. 199.

Route of the Washburn Party, 1870 [map]. Scale not given. Reprinted in: Bonney, Orrin H. and Lorraine G. Bonney. *Battle Drum and Geysers.* 3 vol. Houston: Bonney and Bonney, 1970, vol. 2, folded map, p. 295.

Samuel Lewis' Engraving from Clark's Map [map]. Scale not given. 1814. Reduced and reprinted in: Haines, Aubrey L. *Yellowstone National Park: Its Exploration and Establishment.* Washington, D.C.: U.S. Government Printing Office, 1974, p. 186.

Sketch of the Yellowstone Lake and the Valley of the Upper Yellowstone River [map]. 1:300,000. 1872. In: U.S. Senate. 42nd Congress, 2nd Session. *Report of Reconnaissance of Basin of Upper Yellowstone in 1871* (S. Ex. Doc. 66). Washington, D.C.: U.S. Government Printing Office, 1872. (Serial Set 1479). Reduced and reprinted in: Baldwin, Kenneth. *Enchanted Enclosure: The Army Engineers and Yellowstone National Park.* Washington: U.S. Government Printing Office, 1976, p. 17.

U.S. War Department Map of the Yellowstone and Missouri Rivers and Their Tributaries explored by Capt. W. F. Raynolds Topl. Engrs. and 1st Lt. H. E. Maynadier, 10th Inff, Assistant, 1859-60, to Accompany a Report to the Bureau of Topographical Engineers [map]. Scale not given. In: U.S. Senate. 40th Congress, 1st Session. *Report of Brevet Brigadier General W. F. Raynolds on the Exploration of the Yellowstone and the Country Drained by that River* (S.Ex.Doc. 77). Washington, D.C.: U.S. Government Printing Office, 1868. (Serial Set 1317). Partially reprinted (Yellowstone National Park area) in: Baldwin, Kenneth. *Enchanted Enclosure: The Army Engineers and Yellowstone National Park.* Washington: U.S. Government Printing Office, 1976, p. 7.

Upper Geyser Basin [map]. 1" = 500 ft. In: Ludlow, William. *Report of a Reconnaissance from Carroll, Montana Territory, on the Upper Missouri to the Yellowstone National Park, and Return Made in the Summer of 1875.* Washington: U.S. Government Printing Office, 1876, folded map in pocket. Reprinted in: Ludlow, William.

Exploring Nature's Sanctuary: Captain William Ludlow's Report of a Reconnaissance from Carroll, Montana Territory, on the Upper Missouri to the Yellowstone National Park, and Return Made in the Summer of 1875. Engineer Historical Studies No. 3. Washington: U.S. Government Printing Office, 1985, folded map in pocket.

Upper Geyser Basin, Fire Hole River, Wyoming Territory [map]. Ca 1:18,300. In: Hayden, F. V. "The Hot Springs and Geysers of the Yellowstone and Firehole Rivers." *American Journal of Science and Arts,* 3 (January-June 1872): after page 106; and U.S. House. 42nd Congress, 2nd Session. *Preliminary Report of the United States Geological Survey of Montana and Portions of Adjacent Territories* (H. Ex. Doc. 326). Washington, D.C.: U.S. Government Printing Office, 1872, p. 113. (Serial Set 1520).

William Clark's Manuscript Map [map]. Scale not given. 1806-1811. Reprinted and reduced in: Haines, Aubrey L. *Yellowstone National Park: Its Exploration and Establishment.* Washington, D.C.: U.S. Government Printing Office, 1974, p. 185.

Yellowstone and Grand Teton [map]. 1" = 6.5 mi. In: *National Geographic* 175 (February 1989): 216A [Supplement, Special Places of the World Map].

Yellowstone National Park [map]. Ca 1:6,300. In: Hayden, F. V. "The Yellowstone National Park." *American Journal of Science and Arts,* 3 (January-June 1872): after page 297; and U.S. House. 42nd Congress, 2nd Session. *Preliminary Report of the United States Geological Survey of Montana and Portions of Adjacent Territories* (H.Ex.Doc. 326). Washington, D.C.: U.S. Government Printing Office, 1872, p. 162. (Serial Set 1520).

This book is published for
The Map and Geography Round Table
of the American Library Association
by Speculum Orbis Press
1050 Gage Street
Winnetka, Illinois 60093

The paper is
Pearl White Torchglow Offset
Recycled Stock

Book design after
MAGERT Occasional Paper No. 1

Typesetting was done
in Aldus PageMaker by
Susan Huot
University of Washington

Printed by
The University of Kansas
Printing Services
Lawrence, Kansas